TUMU GONGCHENG ZHUANYE YINGYU

▲ 最新规范
▲ 全国大学版协优秀畅销书

土木工程专业英语（第三版）

主编 田文玉
主审 杨全红

U0240364

重庆大学出版社

内容提要

本书收集了土木工程材料、道路设计、道路施工、桥梁工程、房屋建筑工程、交通工程、工程招标、工程合同、工程监理以及科技文章写作等有关英文资料,目的是让学生通过学习,能够尽可能多地掌握与本专业有关的英文术语,为日后在工作中查阅英文资料打下基础。

本书适合土木工程专业的本科学生学习使用。

图书在版编目(CIP)数据

土木工程专业英语/田文玉主编.--3版.--重庆:重庆大学出版社,2020.7(2023.6重印)
ISBN 978-7-5689-2285-2

Ⅰ.①土… Ⅱ.①田… Ⅲ.①土木工程—英语—高等学校—教材 Ⅳ.①TU

中国版本图书馆 CIP 数据核字(2020)第 111927 号

土木工程专业英语
(第三版)

主编 田文玉
主审 杨全红

策划编辑:曾显跃 鲁 黎

责任编辑:谭 敏　　版式设计:曾显跃
责任校对:邬小梅　　责任印制:张 策

*

重庆大学出版社出版发行
出版人:饶帮华
社址:重庆市沙坪坝区大学城西路21号
邮编:401331
电话:(023) 88617190　88617185(中小学)
传真:(023) 88617186　88617166
网址:http://www.cqup.com.cn
邮箱:fxk@cqup.com.cn(营销中心)
全国新华书店经销
重庆愚人科技有限公司印刷

*

开本:787mm×1092mm　1/16　印张:10.75　字数:350千
2020年7月第3版　　2023年6月第12次印刷
印数:20 501—22 500
ISBN 978-7-5689-2285-2　　定价:39.00元

本书如有印刷、装订等质量问题,本社负责调换
版权所有,请勿擅自翻印和用本书
制作各类出版物及配套用书,违者必究

土木工程专业本科系列教材
编审委员会

主　任　朱彦鹏
副主任　周志祥　程赫明　陈兴冲　黄双华
委　员（按姓氏笔画排序）

于　江	马铭彬	王　旭	王万江	王秀丽
王泽云	王明昌	孔思丽	石元印	田文玉
刘　星	刘德华	孙　俊	朱建国	米海珍
邢世建	吕道馨	宋　彧	肖明葵	沈　凡
杜　葵	陈朝晖	苏祥茂	杨光臣	张东生
张建平	张科强	张祥东	张维全	周水兴
周亦唐	钟　晖	郭荣鑫	黄　勇	黄呈伟
黄林青	彭小芹	程光均	董羽蕙	韩建平
樊　江	魏金成			

第三版前言

《土木工程专业英语》于2005年首次出版,2012年修订后再版。《土木工程专业英语》作为高等学校教材,至今已使用了15年。在第三版中,编者根据教学要求,对 UNIT 10、UNIT 11、UNIT 12、UNIT 13 的内容作了调整、补充,删除了 UNIT 25,其他章节保持不变。

由于编者水平有限,书中难免存在疏漏之处,恳请读者指出,以利于今后改进。

编　者

2020年2月

第一版前言

专业英语是大学英语教学中的一个重要组成部分，是学生从普通英语学习到将英语用到实际工程中的一个有效过渡。本着覆盖面广、知识面宽、信息量大的原则，《土木工程专业英语》充分结合了土建专业的特点，收集了土木工程材料、道路设计、道路施工、桥梁工程、房屋建筑工程、交通工程、工程招标、工程合同、工程监理以及科技文章写作等有关英文资料，目的是让学生通过学习，能够尽可能多地掌握与本专业有关的英文术语，为日后在工作中查阅英文资料打下基础。

《土木工程专业英语》共 24 个 UNITS，每一个 UNIT 包含 Text A 和 Text B 两部分。不同专业方向可根据自身的专业特点选择使用。

全书由田文玉担任主编，杨全红担任主审。其中 UNIT 7 Text B、UNIT 8 Text B、UNIT 9 Text A 由王燕编写，UNIT 1 Text B、UNIT 5 Text B、UNIT 6 Text A 及 Text B、UNIT 7 Text A 由郑智能编写，其余部分由田文玉编写。全书由田文玉负责统稿。

由于编者水平有限，书中难免存在不足之处，恳请读者指出，以利于今后改进。

编 者
2012 年 10 月

Contents

UNIT 1 ··· 1
    Text A   University Courses and Careers in Civil Engineering
    ··· 1
    Text B   The Field in Civil Engineering ······················ 3
UNIT 2 ··· 7
    Text A   Development of Roads ································ 7
    Text B   Return to the Turnpikes ······························ 10
UNIT 3 ··· 12
    Text A   How Is a Modern Road Built ···················· 12
    Text B   Highway Types ·· 14
UNIT 4 ··· 18
    Text A   Road Construction (I) ······························ 18
    Text B   Road Construction (II) ····························· 21
UNIT 5 ··· 24
    Text A   Sight Distance (I) ······································ 24
    Text B   Sight Distance (II) ····································· 27
UNIT 6 ··· 30
    Text A   General Design Criteria of Highway ········ 30
    Text B   Horizontal Alignment ································· 34
UNIT 7 ··· 37
    Text A   Vertical Alignment ····································· 37
    Text B   Medians for Multilane Highways ·············· 39
UNIT 8 ··· 42
    Text A   Design of the Cross Section (I) ············· 42
    Text B   Design of the Cross Section (II) ············ 45
UNIT 9 ··· 48
    Text A   Design of Intersections at Grade ·············· 48
    Text B   Interchanges ·· 51
UNIT 10 ··· 56
    Text A   Portland Cement ·· 56

	Text B	Cement Properties and Storage	60
UNIT 11			65
	Text A	Plain Concrete, Reinforced Concrete, and Prestressed Concrete	65
	Text B	Proportioning, Batching and Mixing of Concrete	69
UNIT 12			73
	Text A	Conveying, Placing, Compacting, and Curing of Concrete	73
	Text B	Concrete Properties	77
UNIT 13			83
	Text A	Durability of Concrete	83
	Text B	Durability of Building Materials	86
UNIT 14			89
	Text A	Rigid Pavement	89
	Text B	Flexible Pavement	90
UNIT 15			95
	Text A	Bituminous Materials	95
	Text B	Bituminous Surface	97
UNIT 16			100
	Text A	Building Materials	100
	Text B	Testing of Materials	105
UNIT 17			108
	Text A	Stress-Strain Relationship of Materials	108
	Text B	Load Classification	110
UNIT 18			113
	Text A	Field Measurement of Density and Moisture Content	113
	Text B	Tests for Determining the Density of Soils	116
UNIT 19			118
	Text A	America on Wheels	118
	Text B	Bus Priorities	120
UNIT 20			122
	Text A	Traffic Engineering (I)	122
	Text B	Traffic Engineering (II)	125
UNIT 21			129
	Text A	Bridges	129

Text B	The Future of Tall Building	132
UNIT 22		136
Text A	Civil Engineering Contracts	136
Text B	Making a Contract	139
UNIT 23		142
Text A	Construction and Building Inspectors	142
Text B	Construction Cost Estimation	146
UNIT 24		153
Text A	Scientific Paper	153
Text B	How to Write a Scientific Paper	155
References		161

UNIT 1

Text A University Courses and Careers in Civil Engineering

Engineering is a profession, which means that an engineer must have a specialized university education. Many government jurisdictions also have licensing procedures which require engineering graduates to pass an examination, similar to the examination for a lawyer, before they can actively start on their careers.

In the university, mathematics, physics, and chemistry are heavily emphasized throughout the engineering curriculum, but particularly in the first two or three years. Mathematics is very important in all branches of engineering, so it is greatly stressed. Today, mathematics includes courses in statistics, which deals with gathering, classifying, and using numerical data, or pieces of information. An important aspect of statistical mathematics is probability, which deals with what may happen when there are different factors, or variables, that can change the results of a problem. Before the construction of a bridge is undertaken, for example, a statistical study is made of the amount of traffic the bridge will be expected to handle[①]. In the design of the bridge, variable such as water pressure on the foundation, impact, the effects of different wind forces, and many other factors must be considered.

Because a great deal of calculation is involved in solving these problems, computer programming is now included in almost all engineering curricula. Computers, of course, can solve many problems involving calculations with greater speed and accuracy than humans. But computers are useless unless they are given clear and accurate instructions and information—in other words, a good program.

In spite of the heavy emphasis on technical subjects in the engineering curriculum, a current trend is to require students to take courses in the social science and the language arts. The relationship between engineering and society is getting closer; it is sufficient, therefore, to say again that the work performed by an engineer affects society in many different and important ways that he or she should be aware of[②]. An engineer also needs a sufficient command of language to be able to prepare reports that are clear and, in many cases, persuasive. An engineer engaged in research will need to be able to write up his or her findings for scientific publications.

The last two years of an engineering program include subjects within the student's field of specialization. For the student who is preparing to become a civil engineer, these specialized courses may deal with such subjects as geodetic surveying, soil mechanics, or hydraulics.

Active recruiting for engineers often begins before the student's last year in the university. Many different corporation and government agencies have competed for the services of engineers in recent years. In the science-oriented society of today, people who have technical training are, of course, in demand. Young engineers may choose to go into environmental or sanitary engineering, for example, where environmental concerns have created many openings; or they may choose construction firms that specialize in highway work; or they may prefer to work with one of the government agencies that deal with water resource. Indeed, the choice is large and varied.

When the young engineer has finally started actual practice, the theoretical knowledge acquired in the university must be applied. He or she will probably be assigned at the beginning to work with a team of engineers. Thus, on-the-job training can be acquired that will demonstrate his or her ability to translate theory into practice to the supervisors.

The civil engineer may work in research, design, construction supervision, maintenance, or even in sales or management. Each of these areas involves different duties, different emphases, and different uses of engineer's knowledge.

Civil engineering projects are almost always unique, each has its own problems and design features. Therefore, careful study is given to each project even before design work begins. The study includes a survey both of topography and subsoil features of the proposed site. It also includes a consideration of possible alternatives, such as a concrete gravity dam or an earth-fill embankment dam. The economic factors involved in each of the possible alternatives must also be weighed. Today, a study usually includes a consideration of the environmental impact of the project. Many engineers, usually working as a team that includes surveyors, specialists in soil mechanics, and experts in design and construction, are involved in making these feasibility studies.

Many civil engineers, among them the top people in the field, work in design. As we have seen, civil engineers work on many different kinds of structures, so it is normal practice for an engineer to specialize in just one kind. In designing buildings, engineers often work as consultants to architectural or construction firms. Dams, bridges, water supply systems, and other large projects ordinarily employ several engineers whose work is coordinated by a system engineer who is in charge of the powerhouse and its equipment. In other cases, civil engineers are assigned to work on a project in another field; in the space program, for instance, civil engineers were necessary in the design and construction of such structures as launching pads and rocket storage facilities.

Construction is a complicated process in almost all engineering projects. It involves scheduling the work and utilizing the equipment and the materials so that costs are kept as low as possible. Safety factor must also be taken into account, since construction can be very dangerous. Many civil engineers therefore specialize in the construction phase.

Words and Expressions

specialized   *a.* 专业的,专门的
jurisdiction   *n.* 管辖权,权限
curriculum   *n.* 课程,学习计划
probability   *n.* 概率
variable   *n.* 变量
persuasive   *a.* 有说服力的
geodetic   *n.* 大地测量学
hydraulics   *n.* 水力学
recruit   *v.* 招聘,征募新人
demonstrate   *v.* 展示,演示
topography   *n.* 地形学
subsoil   *n.* 地基下层土
gravity   *n.* 重力,地心引力
geodetic surveying   大地测量
soil mechanics   土力学
feasibility study   可行性研究
consultant   *n.* 咨询师,顾问
coordinate   *v.* 合作
system engineer   系统工程师
launching pads   (火箭等的)发射台
schedule   *v.* 订计划
construction phase   施工阶段

Notes

①the bridge will be…:为定语从句,省略了关系代词"which"。
②It is sufficient,therefore,to say…:句中,it 为形式主语,to say 为真实主语,that the work…为宾语从句,that he or she…为定语从句,修饰 ways。

Text B   The Field in Civil Engineering

From the Pyramids of Egypt to the Space Station Freedom, civil engineers have always faced the challenges of the future-advancing civilization and building our quality of life.

Today, the world is undergoing tremendous changes—the technological revolution, population growth, environmental concerns, and so on. All create unique challenges for civil engineers of every specialty. The next decades will be the most creative, demanding, and rewarding of times for civil engineers, and now is the best time to find the right career for you.

Today, civil engineers are in the forefront of technology. They are the leading users of sophisticated high-tech products — applying the very latest concepts in computer-aided design (CAD) during design, construction, project scheduling, and cost control.

Civil engineering is about community service, development, and improvement—the planning, design, construction, and operation of facilities essential to modern life, ranging from transit systems to offshore structures to space satellites. Civil engineers are problem solvers, meeting the challenges of pollution, traffic congestion, drinking water and energy needs, urban redevelopment, and community planning.

Our future as a nation will be closely tied to space, energy, the environment, and our ability to interact with and compete in the global economy. You, as a civil engineer, will perform a vital role in linking these themes and improving quality of life for the 21st century. As the technological revolution expands, as the world's population increases, and as environmental concerns mount, your skills will be needed. There is no limit to the personal satisfaction you will feel from helping to make our world a better place to live. Whatever area you choose, be in design, construction, research, teaching, or management, civil engineering offers you a wide range of career choices for your future. Civil engineering is grouped into seven major divisions of engineering: Structural; Environmental; Geotechnical; Water Resources; Transportation; Construction; and Urban Planning. In practice, these are not always hard and fixed categories, but they offer a helpful way to review a very diverse and dynamic field.

Structural

As a structural engineer, you will face the challenge of designing structures that support their own weight and the loads they carry, and that resist wind, temperature, earthquake, and many other forces. Bridges, buildings, offshore structures, space platforms, amusement park rides, and many other kinds of projects are included within this exciting discipline. You will develop the appropriate combination of steel, concrete, and visit the project site to make sure the work is done properly.

Environmental

The skills of environmental engineers are becoming increasingly important as we attempt to protect the fragile resources of our planet. Environmental engineers translate physical, chemical, and biological processes into systems to destroy toxic substances, remove pollutants from water, reduce non-hazardous solid waste volumes, eliminate contaminates from the air, and develop groundwater supplies. In this field, you may be called upon to resolve issues of providing safe drinking water, cleaning up sites contaminated with hazardous materials, disposing of wastewater, and managing solid wastes.

Geotechnical

Geotechnical engineering is required in all aspects of civil engineering, because most projects are supported by the ground. As a geotechnical engineer, you might develop projects below ground, such as tunnels, foundations, and offshore platforms. You will analyze the properties of soil and rock that support and affect the behavior of these structures. You may evaluate the potential settlements of buildings, the stability of slopes and fills, the seepage of ground water and the effects of earthquakes. You will investigate the rocks and soils at a project site and determine the best way to support a structure in the ground. You may also take part in the design and construction of dams, embankments, and retaining walls.

Water Resources

Water is essential to our lives, and as a water resources engineer, you will deal with issues concerning the quality and quantity of water. You will work to prevent floods, to supply water for cities, industry, and irrigation, to treat wastewater, to protect beaches, or to manage and redirect rivers. You might be involved in the design, construction, or maintenance of hydroelectric power facilities, canals, dams, pipelines, pumping stations, locks, or seaport facilities.

Transportation

Because the quality of a community is directly related to the quality of its transportation system, your function as a transportation engineer will be to move people, goods, and materials safely and efficiently. Your challenge will be to find ways to meet our ever increasing travel needs on land, air, and sea. You will design, construct, and maintain all types of transportation facilities, including highways, railroads, airfields, and ports. An important part of transportation engineering is to upgrade our transportation capability by improving traffic control and mass transit systems, and by introducing high-speed trains, people movers, and other new transportation methods.

Construction

As a construction engineer, you are the builder of our future. The construction phase of a project represents the first tangible result of design. Using your technical and management skills will allow you to turn designs into reality—on time and within budget.

You will apply your knowledge of construction methods and equipment, along with the principles of financing, planning, and managing, to turn the designs of other engineers into successful projects.

Urban Planning

As a professional in this area, you will be concerned with the entire development of a community. Analyzing a variety of information will help you coordinate projects, such as projecting street patterns, identifying park and recreation areas, and determining areas for industrial and residential growth. To ensure ready access to your community, coordination with other authorities may be required to integrate freeways, airports, and other related facilities. Successful coordination of a project will require you to be people-oriented as well as technically knowledgeable.

Words and Expressions

forefront   *n.* 最前部,最前线,最活动的中心
structural   *a.* 结构的,建筑的;结构,构造
environmental   *a.* 周围的,环境的
geotechnical   *a.* 岩土的
water resources   水资源
transportation   *n.* 交通,运输
urban planning   城市规划

UNIT 2

Text A Development of Roads

When we speed along a modern highway, we rarely stop to think what it is we are riding on[①]. To understand what a road is, we must study the ways in which people have traveled in the past.

The very first roads were really tracks beaten in the ground by wild animals in prehistoric time. People followed these winding trails because they provided an easy and quick way to get through thick forests. In time, people began to improve the paths by filling holes with earth and laying logs across soft, boggy spots. These attempts were crude, but they were the beginning of road construction.

As people began to transport goods over longer distances, they developed new ways of traveling. First they packed their wares on animals. Then they invented various kinds of sleds. Finally, after the invention of the wheel, they built wagons. Each advancement brought a need for better traveling routes.

Later in history, when well-traveled routes were made sturdier with rocks and stones, the path was raised above the surrounding land, it became a "high way".

The great civilization throughout history were also the great road builders[②]. Roads were necessary to control and extend empires, to permit trade and travel, and to move armies.

Most of these early roads were simply hard-packed dirt, but some were paved with stone blocks or burnt bricks.

The Romans bound their empire together with an extensive system of roads radiating in many directions from Rome. Some of these early roads were of elaborate construction. For example, an Appian Way[③], built southward about 312 BC, illustrates one of the procedures used by the Romans. First a trench was excavated to such a depth that the finished surface would be at ground level. The pavement was placed in three courses: a layer of small broken stones, a layer of small stones with mortar and firmly tamped into place, and a wearing course of massive stone blocks, set and bedded in mortar. Some of the Roman roads are still in existence today. And many modern highways follow the ancient Roman routes.

Few roads were built during the early history of the United States since most of the early settlements were connected with the nearest wharf, but the connecting road usually was just a clearing through the forest. Before the Revolutionary War, travel was mainly on foot or horseback, and roads were merely trails clearing to greater width. Development was extremely slow for a time after the war's end in 1783.

Between 1795 and 1830 numerous turnpikes, particularly in the northeastern states, were built by companies organized to gain profits through toll collections. Few of them were financially successful. During this period many stagecoach lines and freight-hauling companies were organized.

The extension of turnpikes in the United States was abruptly halted by the development of the railroads. In 1830 Peter Cooper constructed America's first steam locomotive, the Tom Thumb, which at once demonstrated its superiority over horse-drawn vehicles. Rapid growth of the railroad for transportation over long distances followed. Cross-country turnpike construction practically ceased, and many already completed fell into disuse. Rural roads served mainly as feeders for the railroads; improvements primarily led to the nearest railroad station and were made largely by local authorities and were to low standards. When it rained, the roads were slippery, and in dry weather they were dusty. People using horses and wagons accepted this. But with the beginning of the 20th century a new invention, the automobile, began to take over the road. The first two decades of the twentieth century saw the improvement of the motor vehicle from a "rich man's toy" to a fairly dependable method for transporting persons and goods. There were strong demands not only from farmers but from bicyclists through the League of American Wheelmen④ for rural road improvement, largely for roads a few miles in length connecting outlying farms with towns and railroad stations. This development has been aptly described as "getting the farmer out of the mud".

From 1920 to 1935, highway development was focused primarily on the completion of a network of all-weather rural roads comparable to the street systems undertaken by local governments. By 1935 highway activities in rural areas have been devoted mainly to an attempt to provide facilities of highway standards and with greater capacity and load-carrying ability. During the same period, increasing attention has been focused on urban areas, which have been struck simultaneously by rapidly increasing population, however population densities resulting from a "flight to the suburbs", and a shift from mass transportation to the private automobile. Indications are that only minor additions to road mileage will be made in the future.

Words and Expressions

track   n. 踪迹,小径
beat tracks   开辟路径
prehistoric   a. 史前的
trail   n. 足迹,小路

boggy　*a.* 多沼泽的
crude　*a.* 粗糙的,不精细的
wares　*n.* (复数)货物,商品
sled　*n.* 雪橇,雪车(以木质或金属长条代替轮子的交通工具)
wagons　*n.* 四轮运货马(或牛)车
sturdy　*a.* 结实的,坚固的
radiate　*v.* 向各方伸展,辐射
elaborate　*a.* 精细的,复杂的
trench　*n.* 沟,沟渠
excavate　*v.* 挖掘
mortar　*n.* 砂浆,胶泥
tamp　*v.* 捣固,夯实
settlement　*n.* 新殖民地,定居点,居民点
bay　*n.* 海湾
wharf　*n.* 码头
clearing　*n.* 开辟出来的空地
turnpike　*n.* (泛指)公路
stagecoach　*n.* 公共马车
freight　*n.* 货物,货运
haul　*v.* 搬运,拖运
locomotive　*n.* 机车,火车头
cross-country　*a.* 越野的
feeder　*n.* 支线
motor-vehicle　机动车,汽车
outlying　*a.* 远离中心的,地处郊区的
apt　*a.* 恰当的,巧妙的
simultaneous　*a.* 同时的,同时发生的
mileage　*n.* 里程,英里数
in time　过了一段时间以后
well-traveled routes　经常行走的路线
hard-packed dirt　压紧或夯实的土
wearing course　磨(耗)损层
toll-collections　征收路税,收取过路费
all-weather roads　晴雨通车路,全天(年)候道路
tamp into place　夯实到位
take over　接管,控制
load-carrying ability　运载能力
mass transportation　公共交通

9

Notes

①… what it is we are riding on… 我们是在什么上面驾车行驶。这是强调句型 it is… that…，句中被强调部分是疑问词 what，关系代词 that 被省去了。例如：You don't know what it is (that) you are doing. It's a mischief!

②本句不宜直译为"历史上伟大的文明也是伟大的道路建设者"，可意译为"历史上伟大的文明时期都是道路的大发展时期"或者"在人类历史上，文明昌盛之时即道路大发展之期"。

③古罗马最著名的大道，建于公元前 312 年，全长 350 多千米，其主干部分保留至今。公元 1784 年，罗马教皇庇护六世修建了新的 Appian Way，由罗马通向 Albano，与旧大道平行。

④the League of American Wheelmen   美国驾车人联合会。

Text B   Return to the Turnpikes

Toll roads, or *turnpikes*, which charge a fee for use, have become important again in the better roads movement. This is the second time they have appeared in America. They were first introduced in Europe during the late 18th century when roads were privately owned and fees collected for using the roads. Poles armed with sharp spikes called pikes stopped the traveler at toll stations. The poles were turned aside after payment, hence the term *turnpike*.

Turnpikes were first introduced into America in 1785. By 1850 there were over 400 in New York State alone. They were the first improved roads in this country. For one type of surface, trunks of tree were placed across the roads and allowed to settle into the roadbed. These were called "corduroy" road.

Some turnpikes were faced with wooden planks. First used in Russia, they were introduced into North America by Sir Charles Edward, governor general of Canada. He built a few miles of plank road near Toronto in 1836. The first plank road in the United States was built in 1845-1846 from Syracuse, N.Y., to Oneida Lake, a distance of about 14 miles.

Most of the turnpikes were between the large cities, such as the Lancaster Turnpike between Philadelphia and Lancaster, Pennsylvania. It was built between 1790 and 1794 and was the first important *macadamized* road in America.

A Scottish engineer, John L. MacAdam (1756-1836), devised the road surface which bears his name. It was made of loosely packed broken stone using water as a binder. Today macadamized surfaces are generally mixed (impregnated) with hot asphalt, tar, or some similar binder. This type of road surface is called *bituminous macadam*.

Turnpike fell into disuse by the middle of the 19th century. The railroad had arrived and proved to be a better means of travel over long distances. Canals too took passenger and freight business from the turnpikes.

The Pennsylvania Turnpike, which was opened in 1940, was the first of the modern turnpikes.

This 359 mile highway extends from the New Jersey border near Philadelphia, Pa., to the Ohio border. World War II interrupted the turnpike movement but at the end of the war a toll road rash began. In 1956 the federal government began a 25-billion-dollar national road-building program. The actual cost of the project, scheduled for completion in 1976, was estimated at 62.5 billion dollars.

Words and Expressions

  toll *n.* 通行税
  turnpike *n.* 征收通行税的路
  fee *n.* 手续费；税
  spike *n.* 大钉，道钉
  trunk *n.* 树干
  corduroy *n.* 木排路，圆木路（沼泽地上用）
  plank *n.* 板（厚2~6英寸，宽9英寸以上）
  governor general 总领事
  macadamize *v.* 铺碎石，用碎石铺路
  loosely *ad.* 松散地
  binder *n.* 黏结剂或胶结料
  impregnate *v.* 使充满，浸透
  asphalt *n.* 沥青
  freight *n.* 货运；水上运输
  border *n.* 边界
  federal *a.* 联邦的
  actual *a.* 实际的
  schedule *v.* 预订，安排，计划
  rash *n.* （突然产生的）一大批

UNIT 3

Text A How Is a Modern Road Built

The first step in building a road is to plan the route. Sometimes the route has been decided by the nature of the land①, but today nature can often be conquered. With powerful modern machinery, whole mountains can be moved and valleys be filled in to make the route as direct as possible.

Then the details are planned: the width of the highway, the number of driving lanes, the number and location of entrances and exits, and the essential strength of the road. All these depend on the amount of traffic that is expected. Modern roads are usually planned for the next 20 years' traffic.

The next step is the testing of the earth foundation on which the road is to rest. Engineers carefully study the soil to learn how solid it is, how much moisture it contains, and how well it drains. Then they decide how the soil should be prepared and packed to provide a good, sturdy foundation, or roadbed. They prescribe the thickness of the road layers, the size of the rocks in them, and the other materials that should be used.

While the testing is under way, a group of surveyors begin to measure the land to find out exactly how much work needs to be done and how much it will cost. Then the construction crew can finally move in.

Giant bulldozers clear the path for the roadbed. They knock over trees and tear large rocks out of the ground. Other powerful earth-moving machines, such as loaders and scrapers, follow in their tracks, scoop up earth and rocks, and dump them into low spots. These filling materials are pressed down tightly with power rollers, and gradually the roadbed becomes a long, level band of hard-packed dirt.

Proper drainage is essential in road building, because if the foundation became soggy, the heavy road would sink into it. And if the water were to freeze in the ground, it would expand and crack the road. To protect the highway from such damage, drainpipes called culverts are laid across the roadbed wherever a strong flow of water is expected. The roadbed itself is shaped so that the middle of the finished road will be higher than the sides. Then water and melting snow will easily drain off its surface into the drainpipes or ditches[②]. The roadbed is given a final grading (smoothing) and is now ready to receive the road itself.

Almost all roads are built in two or more layers, or courses, of rocks or stones. The bottom layer is 10 to 20 centimeters thick and usually made up of larger stones. The upper course has smaller stones and is about 8 centimeters thick. In most roads the lower course is wider so that the edges, or shoulders, of the top course do not break off or sink into soft dirt. After each course is laid, it is compacted by heavy power rollers.

The top, or surface, layer of a road must withstand the weight of heavy vehicles. It must also prevent water from seeping into the roadbed and destroying it. Modern highways are therefore surfaced either with concrete or with bituminous materials, such as asphalt, tar, or heavy oils.

Words and Expressions

conquer　*v.* 征服
machinery　*n.* 机械,机器
lane　*n.* 车道
driving lane　行车道
inside/outside lane　内/外车道
entrance　*n.* 入口
exit　*n.* 出口
moisture　*n.* 湿度,潮湿
drain　*v.* 排水
drainpipe　排水沟管
drainage system　排水系统
pack　*v.* 压紧;夯实;包装
packing course　填层
packing density　夯实密度
prescribe　*v.* 规定,指定;开(药方)
surveyor　*n.* 勘测者,测量员
crew　*n.* 全体队员
construction crew　施工队
bulldozer　*n.* 推土机
loader　*n.* 搬运机,运土机
scraper　*n.* 铲土机
scoop　*v.* 铲起,舀出
dump　*v.* 倾倒
roller　*n.* 压路机

power roller   电动压路机
soggy *a.* 湿润的,湿透的
crack *v.* 使破裂
culvert *n.* 涵洞
ditch *n.* 沟渠
grading *n.* 土工修整
edge *n.* 边缘
shoulder *n.* 路肩
withstand *v.* 抗拒,经得起
seep *v.* 漏,渗
concrete *n.* 混凝土
bituminous *a.* (含)沥青的
asphalt *n.* 地沥青,柏油
tar *n.* 沥青,柏油
be under way   正在进行
knock over   推倒,弄倒
in one's tracks   随即,立刻
filling materials   填料
sink into   陷入

Notes

①nature 不带冠词时意为"大自然",例如:Nature is at its best in spring.大自然在春天最美。但 the nature 指"天性""本性""性质"等,因此文中的句子:Sometimes the routes has been decided by the nature of the land… 应理解为"有时路线由土地的性质确定"。

②drain(*v.*) into… 意为"(水之类的液体)排放到……中",本句可译为"然后水和融雪会顺畅地从路面排入排水管或沟渠"。

Text B　Highway Types

All state highway systems and most of the local highway and street systems encompass several types or classes of highways. At one extreme are high-speed, high-volume facilities carrying through traffic, with no attempt made to serve abutting property or purely local traffic. At the other are local rural roads or streets that carry low volumes, sometimes at low speeds, and with a primary function of land service.

AASHTO[①] Special Committee on Nomenclature gives definitions for various types of highways. Some of these are as follows:

Expressway Divided arterial highway for through traffic with full or partial control of access and generally with grade separations at major intersections.

Freeway Expressway with full control of access.

Parkway Arterial highway for noncommercial traffic, with full or partial control of access, and usually located within a park or a ribbon of parklike developments.

Control of access Condition where the right of owners or occupants of abutting land or other persons to access, light, or view in connection with a highway is fully or partially controlled by public authority.

Full control of access means that the authority to control access is exercised to give preference to through traffic by providing access connections with selected public roads only and by prohibiting crossing at grade of direct private driveway connections.

Partial control of access means that the authority to control access is exercised to give preference to through traffic to a degree that, in addition to access connections with selected public roads, there may be some crossings at grade and some private driveway connections.

The other highway types lack the feature of access control. They include:

Major street or ***major highway*** Arterial highway with intersections at grade and direct access to abutting property, and on which geometric design and traffic-control measures are used to expedite the safe movement of through traffic.

Through street or ***through highway*** Every highway or portion thereof on which vehicular traffic is given preferential right of way, and at the entrances to which vehicular traffic from intersecting highways is required by law to yield right of way to vehicles on such through highway in obedience to either a stop sign or a yield sign, when such signs are erected.

Local road Street or road primarily for access to residence, business, or other abutting property.

It should be pointed out here, however, that the freeway, as typified by the Interstate System, represents the highest type of highway facility, whose advantages include the following:

Capacity On freeways the absence of intersections or crossings at grade and the elimination of marginal friction through access control permit unrestricted, full-time use by moving vehicles, rather than restricted, part-time flow.

Reduced travel time On freeways, time losses from stopping and waiting at intersections are eliminated. In addition, most of the conflicts that contribute to accidents are eliminated, except under unusual circumstances. Drivers normally can and will travel at higher and sustained speeds.

Safer operation On freeways, elimination of conflicts at intersections and along both margins of the roadway and the barring of pedestrians from the right of way usually bring substantial reductions in accidents.

Permanence Access control along freeways prevents the growth of businesses or other activities along the roadway margin. Without access control, these activities generate unordered traffic and parking. In a short time, capacity is reduced and accident potential is substantially increased.

Reduced operation cost, fuel consumption, air pollution, and noise Smoother operations and fewer stops reduce fuel consumption and other operating costs. Reduced fuel consumption in turn reduces air pollution. Smoother operation with fewer stops also greatly reduces noise, particularly that from trucks.

Words and Expressions

encompass　*v.* 包括，围绕

abut　*v.* 邻接，毗连，紧邻

nomenclature　*n.* 术语

expressway　*n.* 快速道路（部分立交）

through traffic　过境交通

grade separation　立体交叉

arterial　*a.* 干线的，主干的，动脉的

intersection　*n.* 交叉（口）

parkway　*n.* 风景区干道

ribbon　*n.*（由市区到郊区）沿干道发展的一系列建筑

preference　*n.* 优先，选择

expedite　*v.* 促进，派遣

obedience　*n.* 服从，顺从

sustained　*a.* 持续的，持久的

pedestrian　*n.* 行人，步行者

permanence　*n.* 永久（性）

state highway　国道

local highway　地方道路

at grade　平面（的）

right of way　通行权，道路用地

yield sign　让路标志

typify　*v.* 作为……的典型，代表

interstate system　州际（道路）系统

marginal friction　路侧摩阻

Note

① AASHTO （American Association of State Highway and Transportation Officials）美国各州公路与运输工作者协会

UNIT 4

Text A Road Construction (I)

Once the road authority has decided to construct a new major road, then it will employ either its own engineers or a consulting engineer to survey the alternative routes and carry out the road design. Information is required, for each of the possible routes, about the detailed ground levels of the terrain, which can now be obtained by aerial photography which is accurate to 6 inch. Details of the types of material for the construction of embankments, and of the geological strata, must be obtained from trial pits and bore holes taken along the line of the route and at bridge sites. The local climatic conditions, such as fog, frost and rain, must also be established. In developed countries, information is required about land values and various environmental factors which may need public enquiries.

From the survey information, the line and level of each of the possible roads will be chosen in accordance with the standards of gradient, sight lines and other factors laid down by the traffic authority. This should minimize the amount of material which has to be excavated and carried to "fill" the adjacent embankments. It is also important to keep the size of bridges needed to cross railways, rivers and other roads to a minimum. Taking into account these various factors, the choice of route is made and the design is carried out.

Arrangements are then made to purchase the land on which the road will run. Detailed drawings, specifications and bills of quantities are prepared so that constructors can tender, normally in competition with each other, for the construction of the work. The consulting engineer or highway authority will usually provide a resident engineer and site staff to ensure that the work is carried out successfully by the constructor in accordance with the drawings and specifications. Within the requirements of the design, the constructor will be responsible for deciding upon the methods of construction to be used.

When the construction team moves on to the site, it is first necessary to clear the line of the new road and fence it where animals may stray on to the works. Trees are cut down, stumps and rocks are grubbed up by bulldozer or where necessary, blasted out by explosives. It may also be necessary to build temporary haul roads① and bridges or fords at the site of the river bridges.

The base of embankments and the slopes of cuttings must be protected from the action of ground water which could cause them to collapse. A primary drainage system is therefore constructed before starting earthworks along the length of the road to cut off the natural ground drainage, and prevent it from entering the works. This is usually done by digging a shallow cut-off ditch with a hydraulic excavator which has a shaped bucket. At the low point of the natural ground, the water flowing in these ditches is taken across the road line in piped or reinforced concrete culverts and allowed to flow away through the existing streams or ditches.

The topsoil is first stripped and stacked ready for spreading on the slopes of cuttings and embankments towards the end of construction. This work is usually done with catepillar tractors towing box scrapers. The main cutting and embankment work is then started, using rubber-tyred scrapers. These are single or twin-engined machines which have a horizontal blade that can be lowered to cut a slice of earth from the ground and collect this earth in the bowl of the scraper. When the scraper bowl is full—some machines can carry up to② 50 cubic yards (38.2 m^3)—the blade is raised and the loaded scraper travels to the "tip" area on the embankment. Particularly in hard digging, it is necessary for the scraper loading operation to be assisted by a pusher bulldozer which pushes the scraper while it is loading to speed up the operation③. For certain types of material, such as chalk④ which may soften in wet weather, or when the excavated material has to be carried for more than two miles (3 km), the excavation may be done using face shovel excavators loading into dump trucks. When rock is encountered, it is first shattered with explosives and then loaded by face shovel.

At the embankment, the earth is spread by the scrapers into a thin layer about 12 in. (305 mm) thick which is leveled by bulldozers and then compacted by caterpillar tractors towing rollers, or by self-propelled rollers which work more quickly. It is vital that the successive embankment layers are properly compacted so that the embankment is stable. The inevitable earth settlement will also be kept to a minimum to prevent damage or excessive maintenance to the finished road.

On completion of the earthworks, further shallow drain trenches, about 4 ft (1.2 m) deep, are constructed to keep the top layer of cutting or embankment free from water, which would weaken it. Pipes are laid in those trenches, which are then filled with gravel. In or adjacent to these trenches are laid further pipes to carry the water collected in road gullies from the finished road surface⑤.

Words and Expressions

survey   *v.* 勘探,踏勘
terrain   *n.* 地域,领地
aerial   *a.* 空中的

stratum　*n.*［复］strata　岩层,地层;薄片;阶层
geological stratum　地质层
gravel stratum　砂砾层
pit　*n.* 坑,槽
trial pit　探井,试坑
bore　*v.* 挖坑
gradient　*n.* 坡度,倾斜度;梯度,陡度
purchase　*v.* 购买
bills of quantity　费用
tender　*v.*（常与 for 连用）投标（做某事）,履行（某契约）
consulting engineer　顾问工程师
resident engineer　工地工程师,驻地工程师
stray　*v.* 离群,走离;走失;迷失;闲逛
stump　*n.* 树桩,残根
grub　*n.* 挖,掘;(常与 up,out 连用)连根挖出
bulldoze　*v.*（用推土机）清除,平整,挖出
bulldozer　*n.* 推土机,(推土机前的)推土刀
blast　*v.* 爆炸,炸开
explosive　*n.* 炸药,爆炸物
ford　*n.* 过水路面
cutting　*n.* 挖方
collapse　*v.* 坍塌
earthworks　*n.* 土方(工程)
hydraulicity　*n.* 水凝性(水泥)
hydraulic excavator　水力冲泥机
bucket　*n.* 铲斗,勺斗
culvert　*n.* 涵洞,管道,阴沟
strip　*v.* 剥去,除掉,脱
stack　*v.* 堆叠
caterpillar　*n.* 履带,坦克车,战车
caterpillar tractor　履带式拖拉机
tow　*v.* 拖,牵引
scraper　*n.* 平土机,铲土机,电耙
box scraper　箱形电耙
blade　*n.* 刀片,叶片
slice　*n.* 薄片,一份,部分
shovel　*n.* 铲,挖掘机

face shovel　正铲挖土机
excavator-type shovel　挖掘机
dump　*v.* 倾倒（垃圾）
encounter　*v.* 遇到，相遇
shatter　*v.* 破碎
level　*v.* 使……平坦，平等；夷平，摧毁
inevitable　*a.* 不可避免的，必然的
settlement　*n.* 沉降，陷落，下沉
trench　*n.* 沟渠，管沟
gully　*n.* 冲沟，集水沟
in accordance with　根据，按照
sight lines　视线，瞄准线
take... into account/consideration　考虑
in competition with　与……竞争
decide upon　决定
single or twin-engined machines　单发动机或双发动机机器
speed up　加快，加速
earth settlement　土方沉降

Notes

①temporary haul roads 指筑路时为运输材料而修的临时运输便道。

②up to sth.意为"多达""直到"，例如：Research suggests that up to half of those who were prescribed the drug have suffered side effects.研究表明，在那些服用药物的人当中，有多达一半的人会受到副作用影响。

③while it is loading 是时间状语从句。to speed up...是目的状语；本句可译为：特别是在挖掘硬土时，铲土机在操作时还需用一辆后推式推土机将其顶住以加速操作。

④chalk 意指白垩，是一种松软的方解石粉块，$CaCO_3$，含有不等量的硅石、石英、长石或其他矿物杂质，通常为灰白色或黄白色，从贝类化石中制得。

⑤这是一个倒装句，介词短语作状语置于句首引起倒装，主语是 further pipes to carry the water collected in road gullies from the finished road surface。in 和 adjacent to 共用了宾语 these trenches。本句可译为：在路边的沟里或附近铺设排水管，以排除从路面的水沟中汇集的水。

Text B　Road Construction（Ⅱ）

It is necessary to phase the bridge construction periods so that the bridges are completed ahead of the paving operations and the existing roads diverted over or through them. The carriageway (pavement) paving operation then begins at the top layer of earth—the formation—being accurately trimmed to a 2 inch (51 mm) tolerance by scrapers or a grader. A grader is a wheeled machine which has a steel blade mounted horizontally between its four wheels. This blade can be accurately raised, lowered or tilted by the driver to cut a precise surface.

A sub-base layer of gravel or crushed rock, generally 12 inch (305 mm) thick, is then spread over the formation to increase its load bearing capacity. Alternatively, when the earth is suitable and imported materials are difficult to obtain, it may be possible to mix a quantity of dry cement into the top layer of earth, which is then damped with water to cause it to harden. This is called *soil stabilization*.

The final layers of the road are then built, either of concrete or of tar-bitumen *black-top* materials.

The bitumen and stone are heated and mixed together in a site mixing plant, and brought hot, by truck, to the laying point. The material is then tipped into a *paver* which spreads it evenly to a thickness of 2.5 inch (64 mm) and to an accurate level. This layer is compacted by road rollers to give a firm surface. The accuracy of each successive layer until the wearing course (usually of asphalt) provides the accuracy of the finished road. Bitumen coated stone chippings are spread over the top surface and rolled into it while they are still hot.

If the final surface is to be concrete, then this will consist of a concrete slab approximately 10 inch (254 mm) thick. The actual thickness will depend upon whether the concrete is reinforced or not. Joints will be incorporated in this slab at about 15 ft (4.6 m) intervals to enable expansion and contraction of the concrete to take place.

Conventionally, the concrete is laid between temporary steel road forms, which support the edge of the concrete slab, by a concrete train. This consists of a series of machines which run on rails supported on the road forms. The forms and thus the rail are accurately laid to level well ahead of the train and provide the level control for the finished road surface. The first machine in the train is a placer spreader which puts the concrete, transported by truck from the concrete mixing plant, between the road forms. The concrete is then compacted and trimmed to true level by successive machines. To provide a skid resistant surface, the wet concrete is then lightly brushed or otherwise grooved to a shallow depth.

Recently slip form machines have been developed, and by using these machines to form the concrete slab, it is possible to eliminate the lengthy process of accurately laying out road forms. These slip form pavers incorporate traveling side forms within the body of the machine. The degree of vibration compacting the concrete is much greater than with the conventional train so that after the moving forms—approximately 15 ft (5 m) long—have slipped past, the fresh concrete is able to stand up without further support. The surface level of the finished concrete is formed by the same machine which is controlled, both for level and direction, by means of electronic or hydraulic sensor controls which follow string lines placed at each side of the machine along the carriageway. With this paver, it is possible to achieve up to 6 ft (1.8 m) per minute.

Words and Expressions

phase   *v.* 使……按计划进行
carriageway   *n.* 路面
tolerance   *n.* 公差
grader   *n.* 平路机,推土机
tilt   *v.* 倾斜
soil stabilization   稳定土
tip   *v.* 倒,倾泻
paver   *n.* 摊铺机
chippings   *n.* 屑
slab   *n.* 厚板
joint   *n.* 连接杆
incorporate   *v.* 结合
interval   *n.* 间隔
expansion   *n.* 膨胀
contraction   *n.* 收缩
conventionally   *ad.* 通常
placer spreader   铺料机
brush   *v.* 刷
groove   *v.* 扫
vibration   *n.* 振动
approximately   *ad.* 大约

UNIT 5

Text A   Sight Distance (I)

For safe vehicle operation, highway must be designed to give drivers a sufficient distance of clear vision ahead so that they can avoid unexpected obstacles and can pass slower vehicles without danger. Sight distance is the length of highway visible ahead to the driver of a vehicle. The concept of safe sight distance has two facets: "stopping" (or "nonpassing") and "passing".

At times large objects may drop onto a roadway and will do serious damage to a motor vehicle that strikes them. Again a car or truck may be forced to stop in the traffic lane in the path of following vehicles. In either instance, proper design requires that such hazards become visible at distances great enough that drivers can stop before hitting them. Furthermore, it is unsafe to assume that one oncoming vehicle may avoid trouble by leaving the lane in which it is traveling, for this might result in loss of control or collision with another vehicle[①].

Stopping sight distance is made up of two elements. The first is the distance traveled after the obstruction comes into view but before the driver applies his brakes. During this period of perception and reaction, the vehicle travels at its initial velocity. The second distance is consumed while the driver brakes the vehicle to a stop. The first of these two distances is dependent on the speed of the vehicle and the perception time and brake-reaction time of the operator. The second distance depends on the speed of the vehicle; the condition of brakes, tires, and roadway surface; and the alignment and grade of the highway.

On two-lane highway, opportunity to pass slow-moving vehicles must be provided at intervals. Otherwise capacity decreases and accidents increase as impatient drivers risk head-on collisions by passing when it is unsafe to do so. The minimum distance ahead that must be clear to permit safe passing is called the passing sight distance.

In deciding whether or not to pass another vehicle, the driver must weigh the clear distance available to him against the distance required to carry out the sequence of events that make up the passing maneuver. Among the factors that will influence his decision are the degree of caution that he exercises and the accelerating ability of his vehicle[②]. Because humans differ markedly, passing practices, which depend largely on human judgment and behavior rather than on the laws of mechanics, vary considerably among drivers. To establish design values for passing sight distances, engineers observed the passing practices of many drivers. Passing sight distance standards are established based on basic observations which were made during the period 1938-1941. Assumed operating conditions are as follows:

1) The overtaken vehicle travels at a uniform speed.

2) The passing vehicle has reduced speed and trails the overtaken one as it enters the passing section.

3) When the passing section is reached, the driver requires a short period of time to perceive the clear passing section and to react to start his maneuver.

4) Passing is accomplished under what may be termed a delayed start and a hurried return in the face of opposing traffic. The passing vehicle accelerates during the maneuver and its average speed during occupancy of the left lane is 10 mph higher than that of the overtaken vehicle.

5) When the passing vehicle returns to its lane there is a suitable clearance length between it and an oncoming vehicle in the other lane.

The five distances, in sum, make up passing sight distance.

Words and Expressions

sight distance 视距
operation *n.* 操作,作用
facet *n.* 面,某一方面
stopping *n.* 停车
nonpassing 禁止超车,停车
passing *a.* 超车
strike *v.* 撞击,冲击
hazard *n.* 危险
oncoming *a.* 迎面而来,接近
collision *n.* 碰撞,冲突
obstruction *n.* 阻挡,阻碍
apply *v.* 应用,使用,适用
brake *n.* 制动器,刹车; *v.* 阻碍
perception *n.* 感觉
initial *a.* & *v.* & *n.* 最初,开始

velocity　*n.* 速度
consume　*v.* 消耗
alignment　*n.* (道路)线形
grade　*n.* 坡度
carry out　进行,实现,执行
sequence　*n.* 连续
event　*n.* 事件,事情
make up　组成
maneuver　*n. & v.* (调动车辆的)机动动作
accelerate　*v. & n.* 加速,促进
markedly　*ad.* 显著地
mechanics　*n.* [复]力学
establish　*v.* 建立,规定,安置
overtake　*v.* 追上,超过
trail　*v. & n.* 跟踪,痕迹
perceive　*v.* 察觉,看见
accomplish　*v.* 完成
delay　*v.* 延误,推迟
hurry　*n. & v.* 匆忙
in the face of　面对
occupancy　*n.* 占有,占用
clearance　*n.* 净空

Notes

①Furthermore, it is unsafe to assume that the oncoming vehicle may avoid trouble by leaving the lane in which it is traveling, for this might result in loss of control or collision with another vehicle. 而且,认为迎面(对着障碍)开来的车辆以离开它所行驶的车道来避开障碍是不安全的,因为这可能会失去控制或与其他车辆相撞。

②Among the factors that will influence his decision are the degree of caution that he exercises and the accelerating ability of his vehicle. 影响驾驶员决策的因素有二:一是他开车的小心程度;二是车辆的加速性能。

本句主语是 degree 和 ability,are 是谓语动词,that will influence his decision 是定语从句,其先行词是 factors,that he exercises 也是定语从句,其先行词是 degree。

Text B Sight Distance(Ⅱ)

General

Sight distance is the continuous length of highway ahead visible to the driver. In design, two sight distances are considered: passing sight distance and stopping sight distance. Stopping sight distance is the minimum sight distance to be provided at all points on multi-lane highways and on two-lane roads when passing sight distance is not economically obtainable.

Stopping sight distance also is to be provided for all elements of interchanges and intersections at grade, including driveways.

Tab. 1 shows the standards for passing and stopping sight distance related to design speed.

Tab. 1 Sight distances for design

Design Speed(m/h)	Sight Distance in feet	
	Stopping* Minimum	Passing* Minimum
25	155	900
30	200	1090
35	250	1280
40	305	1470
45	360	1625
50	425	1835
55	495	1985
60	570	2135
65	645	2285
70	730	2480

* Not applicable to multi-lane highways.

Passing Sight Distance

Passing sight distance is the minimum sight distance that must be available to enable the driver of one vehicle to pass another vehicle, safely and comfortably, without interfering with the speed of an oncoming vehicle traveling at the design speed, should it come into view after the overtaking maneuver is started. The sight distance available for passing at any place is the longest distance at which a driver whose eyes are 3.5 feet above the pavement surface can see the top of an object 3.5 feet high on the road.

Passing sight distance is considered only on two-lane roads. At critical locations, a stretch of four-lane construction with stopping sight distance is sometimes more economical than two lanes with passing sight distance.

Stopping Sight Distance

The minimum stopping sight distance is the distance required by the driver of a vehicle, traveling at a given speed, to bring his vehicle to a stop after an object on the road becomes visible. Stopping sight distance is measured from the driver's eyes, which is 3.5 feet above the pavement surface, to an object 2 feet high on the road.

The stopping sight distance should be increased when sustained downgrades are steeper than 3 percent. Increases in the stopping sight distances on downgrades are indicated in A Policy on Geometric Design of Highways and Streets, AASHTO, 2001.

Stopping Sight Distance on Horizontal Curves

Where an object off the pavement such as a longitudinal barrier, bridge pier, bridge rail, building, cut slope, or natural growth restricts sight distance, the minimum radius of curvature is determined by the stopping sight distance.

For sight distance calculations, the driver's eyes are 3.5 feet above the center of the inside lane (inside with respect to curve) and the object is 2 feet high. The line of sight is assumed to intercept the view obstruction at the midpoint of the sight line and 2.75 feet above the center of the inside lane. Of course, the midpoint elevation will be higher or lower than 2.75 feet, if it is located on a sag or crest vertical curve respectively. The clear distance (M) is measured from the center of the inside lane to the obstruction.

The general problem is to determine the clear distance from the centerline of inside lane to a median barrier, retaining wall, bridge pier, abutment, cut slope, or other obstruction for a given design speed. Radius of curvature and sight distance for the design speed determine the middle ordinate (M) which is the clear distance from centerline of inside lane to the obstruction. When the design speed and the clear distance to a fixed obstruction are known, this figure also gives the required minimum radius which satisfies these conditions.

When the required stopping sight distance would not be available because of an obstruction such as a railing or a longitudinal barrier, the following alternatives shall be considered: increase the offset to the obstruction, increase the horizontal radius, or do a combination of both. However, any alternative selected should not require the width of the shoulder on the inside of the curve to exceed 12 feet, because the potential exists that motorists will use the shoulder in excess of that width as a passing or travel lane.

When determining the required middle ordinate (M) distance on ramps, the location of the driver's eye is assumed to be positioned 6 feet from the inside edge of pavement on horizontal curves.

The designer is cautioned in using the values from Fig. 1 since the stopping sight distances and middle ordinates are based upon passenger vehicles. The average driver's eye height in large trucks is approximately 120 percent higher than a driver's eye height in a passenger vehicle. However, the required minimum stopping sight distance can be as much as 50 percent greater than the distance required for passenger vehicles. On routes with high percentages (10 percent or more) of truck traffic, the designer should consider providing greater horizontal clearances to vertical sight obstructions to accommodate the greater stopping distances required by large trucks. The approximate middle ordinate (M) required for trucks is 2.5 times the value obtained from Fig. 1 for passenger vehicles.

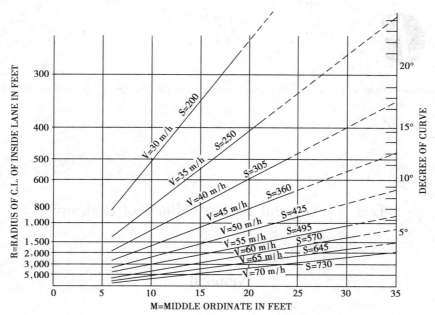

Fig. 1 Minimum stopping sight distance on horizontal curves

Words and Expressions

passing sight distance   超车视距
stopping sight distance   停车视距

UNIT 6

Text A   General Design Criteria of Highway

General

Geometric design is the design of the visible dimensions of a highway with the objective of forming or shaping the facility to the characteristics and behavior of drivers, vehicles and traffic. Therefore, geometric design deals with features of location, alignment, profile, cross section, intersection and highway types.

Highway Classification

Highway classification refers to a process by which roadways are classified into a set of sub-systems, described below, based on the way each roadway is used. Central to this process is an understanding that travel rarely involves movement along a single roadway. Rather each trip or sub-trip initiates at a land use, proceeds through a sequence of streets, roads and highways, and terminates at a second land use.

The highway classification process is required by federal law. Each state must assign roadways into different classes in accordance with standards and procedures established by the Federal Highway Administration. Separate standards and procedures have been established for rural and urban areas. For a further description of the classification process, see Highway Functional Classification: Concepts, Criteria and Procedures, FHWA[1], revised March 1989.

Design Controls

The location and geometric design of highways are affected by numerous factors and controlling features. These may be considered in two broad categories as follows:
(1) Primary Controls
    a. Highway Classification   b. Topography and Physical Features   c. Traffic
(2) Secondary Controls
    a. Design Speed   b. Design Vehicle   c. Capacity

Primary Controls

Highway Classification

Separate design standards are appropriate for different classes of roads, since the classes serve different types of trips and operate under different conditions of both speed and traffic volume. The design of streets and highways on the State highway system should conform to the guidelines as indicated in specifications for highway. In special cases of restrictive or unusual conditions, it may not be practical to meet these guide values.

Topography and Physical Features

The location and the geometric features of a highway are influenced to a large degree by the topography, physical features, and land use of the area traversed. The character of the terrain has a pronounced effect upon the longitudinal features of the highway, and frequently upon the cross sectional features as well. Geological conditions may also affect the location and the geometrics of the highway. Climatic, soil and drainage conditions may affect the profile of a road relative to existing ground.

Man-made features and land use may also have considerable effect upon the location and the design of the highway. Industrial, commercial, and residential areas will each dictate different geometric requirements.

Traffic

The traffic characteristics, volume, composition and speed, indicate the service for which the highway improvement is being made and directly affects the geometric features of design.

The traffic volume affects the capacity, and thus the number of lanes required. For planning and design purposes, the demand of traffic is generally expressed in terms of the design-hourly volume (DHV), predicated on the design year. The design year for new construction and reconstruction is to be 20 years beyond the anticipated date of Plans, Specifications and Estimate (PS & E), and 10 years beyond the anticipated date of PS & E for resurfacing, restoration and rehabilitation projects.

The composition of traffic, i.e., proportion of trucks and buses, is another characteristic which affects the location and geometrics of highways. Types, sizes and loadpower characteristics are some of the aspects taken into account.

Secondary Controls

Design Speed

"Design Speed" is a selected speed used to determine the various design features of the roadway. The assumed design speed should be a logical one with respect to topography, anticipated operating speed, the adjacent land use, and the functional classification of the highway. Except for local streets where speed controls are frequently included intentionally, every effort should be made to use as high a design speed as practicable to attain a desired degree of safety, mobility and efficiency within the constraints of environmental quality, economics, aesthetics and social or political impacts. Once the design speed is selected, all of the pertinent features of the highway should be related to it to obtain a balanced design. Above minimum design values should be used, where practical. Some design features, such as curvature, superelevation, and sight distance are directly related to and vary appreciably with, design speed. Other features, such as widths of lanes and shoulders, and clearances to walls and rails, are not directly related to design speed, but they affect vehicle speeds. Therefore, wider lanes, shoulders, and clearances should be considered for higher design speeds. Thus, when a changes is made in design speed, many elements of the highway design will change accordingly.

Since design speed is predicated on the favorable conditions of climate and little or no traffic on the highway, it is influenced principally by:

Character of the terrain; Extent of manmade features; Economic considerations (as related to construction and right-of-way costs).

These three factors apply only to the selection of a specific design speed within a logical range pertinent to a particular system or classification of which the facility is a part.

Design Vehicle

The physical characteristics of vehicles and the proportions of the various size vehicles using the highways are positive controls in geometric design. A design vehicle is a selected motor vehicle, the weight, dimensions and operating characteristics of which are used to establish highway design controls to accommodate vehicles of a designated type.

Capacity

The term "capacity" is used to express the maximum number of vehicles which have a reasonable expectation of passing over a section of a lane or a roadway during a given time period under prevailing roadway and traffic conditions. However, in a broad sense, capacity encompasses the relationship between highway characteristics and conditions, traffic composition and flow patterns, and the relative degree of congestion at various traffic volumes throughout the range from light volumes to those equaling the capacity of the facility as defined above.

Highway capacity information serves three general purposes:

1) For transportation planning studies to assess the adequacy or sufficiency of existing highway networks to current traffic demand, and to estimate when, in time, projected traffic demand, may exceed the capacity of the existing highway network or may cause undesirable congestion on the highway system.

2) For identifying and analyzing bottleneck locations (both existing and potential), and for the evaluation of traffic operational improvement projects on the highway network.

3) For highway design purposes.

Level of Service

The level of service concept places various traffic flow conditions into 6 levels of service. These levels of service, designated A through F, from best to worst, cover the entire range of traffic operations that may occur.

The factors that may be considered in evaluating level of service include the following:

(1) Speed and travel time;

(2) Traffic interruptions or restrictions;

(3) Freedom to maneuver;

(4) Safety;

(5) Driving comfort and convenience;

(6) Economy.

Words and Expressions

geometric design 线形(几何)设计
alignment *n.* & *v.* 线形(尤其指道路中线的位置与方向);定线
profile *n.* 纵断面,剖面
cross section 横断面
intersection *n.* 交叉,交点,(道路)平面交叉口
topography *n.* 地形,地形学
design vehicle 设计车辆
terrain *n.* 地形,地势
location *n.* 定线,定位
capacity *n.* 通行能力
level of service 服务水平

Note

①FHWA: Federal Highway Administration　联邦公路管理局

Text B　Horizontal Alignment

General

In the design of horizontal curves, it is necessary to establish the proper relationship between design speed, curvature and superelevation. Horizontal alignment must afford at least the minimum stopping sight distance for the design speed at all points on the roadway.

The major considerations in horizontal alignment design are: safety, grade, type of facility, design speed, topography and construction cost. In design, safety is always considered, either directly or indirectly. Topography controls both curve radius and design speed to a large extent. The design speed, in turn, controls sight distance, but sight distance must be considered concurrently with topography because it often demands a larger radius than the design speed. All these factors must be balanced to produce an alignment that is safe, economical, in harmony with the natural contour of the land and, at the same time, adequate for the design classification of the roadway or highway.

Superelevation

When a vehicle travels on a horizontal curve, it is forced radially outward by centrifugal force. This effect becomes more pronounced as the radius of the curve is shortened. This is counterbalanced by providing roadway superelevation and by the side friction between the vehicle tires and the surfacing. Safe travel at different speeds depends upon the radius of curvature, the side friction, and the rate of superelevation.

When the standard superelevation for a horizontal curve cannot be met, a design exception will be required. However, the highest practical superelevation should be selected for the horizontal curve design.

The minimum superelevation to be used is 1.5 percent on flat radius curves requiring superelevation ranging from 1.5 percent to 2.0 percent, the superelevation should be increased by 0.5 percent in each successive pair of lanes on the low side of the superelevation when more than two lanes are superelevated in the same direction.

It may be appropriate to provide adverse crown on flat radius curves (less than 2 percent superelevation) to avoid water buildup on the low side of the superelevation when there are more than three lanes draining across the pavement (This design treatment would require a design exception). Another option is to construct a permeable surface course or a high macrotexture surface course since these surfaces appear to have the highest potential for reducing hydroplaning accidents. Also, grooving the pavement perpendicular to the traveled way may be considered as a corrective measure for severe localized hydroplaning problems.

(1) Axis of Rotation

For undivided highways, the axis of rotation for superelevation is usually the centerline of the traveled way. However, in special cases where curves are preceded by long, relatively level tangents, the plane of superelevation may be rotated about the inside edge of the pavement to improve perception of the curve. In flat terrain, drainage pockets caused by superelevation may be avoided by changing the axis of rotation from the centerline to the inside edge of the pavement.

(2) Superelevation Transition

The superelevation transition consists of the superelevation runoff (length of roadway needed to accomplish the change in outside-lane cross slope from zero to full superelevation or vice versa) and tangent runout (length of roadway needed to accomplish the change in outside-lane cross slope from the normal cross slope to zero or vice versa). The definition of and method of deriving superelevation runoff and runout in this manual is the same as described in the AASHTO publication A Policy on Geometric Design of Highways and Streets, 2001.

The superelevation transition should be designed to satisfy the requirements of safety and comfort and be pleasing in appearance.

Curvature

The changes in direction along a highway are basically accounted for by simple curves or compound curves. Excessive curvature or poor combinations of curvature generate accidents, limit capacity, cause economic losses in time and operating costs, and detract from a pleasing appearance. To avoid these poor design practices, the following general controls should be used.

(1) Curve Radii for Horizontal Curves

For specific design speeds Standards gives the minimum radius of open highway curves. Every effort should be made to exceed the minimum values. Minimum radii should be used only when the cost or other adverse effects of realizing a higher standard are inconsistent with the benefits. Where a longitudinal barrier is provided in the median, the above minimum radii may need to be increased or the adjacent shoulder widened to provide adequate horizontal stopping sight distance.

The suggested minimum radius for a freeway is 3,000 feet in rural areas and 1,600 feet in urban areas. For a land service highway, the preferred minimum radius is 1,600 feet and 1,000 feet for design speeds of 60 mph and 50 mph respectively.

Due to the higher center of gravity on large trucks, sharp curves on open highways may contribute to truck overturning. Overturning becomes critical on radii below approximately 700 feet. Where new or reconstructed curves on open highways with radii less than 700 feet must be provided, the design of these radii shall be based upon a design speed of at least 10 mph greater than the anticipated posted speed.

(2) Alignment Consistency

Sudden reductions in standards introduce the element of surprise to the driver and should be avoided. Where physical restrictions on curve radius cannot be overcome and it becomes necessary to introduce curvature of a lower standard than the design speed for the project, the design speed between successive curves shall change not more than 10 mph. Introduction of a curve for a design speed lower than the design speed of the project shall be avoided at the end of a long tangent or at other locations where high approach speeds may be anticipated.

(3) Stopping Sight Distance

Horizontal alignment should afford at least the desirable stopping sight distance for the design speed at all points of the highway. Where social, environmental or economic impacts do not permit the use of desirable values, lesser stopping sight distances may be used, but shall not be less than the minimum values.

(4) Curve Length and Central Angle

The following is applicable for freeways and rural arterial highways. Desirably, the minimum curve length for central angles less than 5 degrees should be 500 feet long, and the minimum length should be increased 100 feet for each 1 degree decrease in the central angle to avoid the appearance of a kink. For central angles smaller than 30 minutes, no curve is required. In no event shall sight distance or other safety considerations be sacrificed to meet the above requirement.

(5) Reversing Curves

The intervening tangent distance between reverse curves should, as a minimum, be sufficient to accommodate the superelevation transition. For design speeds of 50 mph and greater, longer tangent lengths are desirable.

(6) Broken Back Curves

A broken back curve consists of two curves in the same direction joined by a short tangent. Broken back curves are unsightly and violate driver expectancy. A reasonable additional expenditure may be warranted to avoid such curvature.

The intervening tangent distance between broken back curves should, as a minimum, be sufficient to accommodate the superelevation transition. For design speeds of 50 mph and greater, longer tangent lengths are desirable.

(7) Alignment at Bridges

Superelevation transitions on bridges almost always result in an unsightly appearance of the bridge and the bridge railing. Therefore, if at all possible, horizontal curves should begin and end a sufficient distance from the bridge so that no part of the superelevation transition extends onto the bridge. Alignment and safety considerations, however, are paramount and shall not be sacrificed to meet the above criteria.

Words and Expressions

horizontal curves 平曲线
superelevation *n.* 超高
side friction 侧向摩擦
superelevation transition 超高过渡
alignment consistency 线形的一致性

UNIT 7

Text A Vertical Alignment

General

The profile line is a reference line by which the elevation of the pavement and other features of the highway are established. It is controlled mainly by topography, type of highway, horizontal alignment, safety, sight distance, construction costs, cultural development, drainage and pleasing appearance. The performance of heavy vehicles on a grade must also be considered. All portions of the profile line must meet sight distance requirements for the design speed of the road.

In flat terrain, the elevation of the profile line is often controlled by drainage considerations. In rolling terrain, some undulation in the profile line is often advantageous, both from the standpoint of truck operation and construction economy. But, this should be done with appearance in mind; for example, a profile on tangent alignment exhibiting a series of humps visible for some distance ahead should be avoided whenever possible. In rolling terrain, however, the profile usually is closely dependent upon physical controls.

Separate Grade Lines

Separate or independent profile lines are appropriate in some cases for freeways and divided arterial highways.

They are not normally considered appropriate where medians are less than 30 feet. Exceptions to this may be minor differences between opposing grade lines in special situations.

In addition, appreciable grade differentials between roadbeds should be avoided in the vicinity of at-grade intersections. For traffic entering from the crossroad, confusion and wrong-way movements could result if the pavement of the far roadway is obscured due to an excessive differential.

Standards for Grade

The minimum grade rate for freeways and land service highways with a curbed or bermed section is 0.3 percent. On highways with an umbrella section, grades flatter than 0.3 percent may be used where the shoulder width is 8 feet or greater and the shoulder cross slope is 4 percent or greater.

For maximum grades for urban and rural land service highways and freeways, see Tab. 1.

Tab. 1　Maximum grades(%)

Rural Land Service Highways							
Type of Terrain	Design Speed (m/h)						
	30	40	45	50	55	60	65
Level	—	5	5	4	4	3	3
Rolling	—	6	6	5	5	4	4
Mountainous	—	8	7	7	6	6	5

Vertical Curves

Properly designed vertical curves should provide adequate sight distance, safety, comfortable driving, good drainage, and pleasing appearance. On new alignments or major reconstruction projects on existing highways, the designer should, where practical, provide the desirable vertical curve lengths. The use of minimum vertical curve lengths should be limited to existing highways and those locations where the desirable values or values greater than the minimum would involve significant social, environmental or economic impacts.

Heavy Grades

Except on level terrain, often it is not economically feasible to design a profile that will allow uniform operating speeds for all classes of vehicles. Sometimes, a long sustained gradient is unavoidable.

From a truck operation standpoint, a profile with sections of maximum gradient broken by length of flatter grade is preferable to a long sustained grade only slightly below the maximum allowable. It is considered good practice to use the steeper rates at the bottom of the grade, thus developing slack for lighter gradient at the top or elsewhere on the grade.

Coordination with Horizontal Alignment

A proper balance between curvature and grades should be sought. When possible, vertical curves should be superimposed on horizontal curves. This reduces the number of sight distance restrictions on the project, makes changes in profile less apparent, particularly in rolling terrain, and results in a pleasing appearance. For safety reasons, the horizontal curve should lead the vertical curve. On the other hand, where the change in horizontal alignment at a grade summit is slight, it safely may be concealed by making the vertical curve overlay the horizontal curve.

When vertical and horizontal curves are thus superimposed, the superelevation may cause distortion in the outer pavement edges. Profiles of the pavement edge should be plotted and smooth curves introduced to remove any irregularities.

A sharp horizontal curve should not be introduced at or near a pronounced summit or grade sag. This presents a distorted appearance and is particularly hazardous at night.

Climbing Lane

A climbing lane is an auxiliary lane introduced at the beginning of a sustained positive grade for the diversion of slow traffic.

Generally, climbing lanes will be provided when the conditions in Text B are satisfied. These conditions could be waived if slower moving truck traffic was the major contributing factor causing a high accident rate and could be corrected by addition of a climbing lane.

Words and Expressions

elevation   *n.* 高程
level   *terrain.* 平原
rolling   *terrain.* 微丘
mountainous   *terrain.* 山区
climbing lane   爬坡车道

Text B   Medians for Multilane Highways

Positive separation between opposing streams of traffic has proved to be effective means for reducing headlight glare, conflicts, and accidents on multilane highways. Today medians in some form are an absolute requirement for all free highways. At intersections where roads or streets cross expressways or major city streets, medians, if wide enough, provide further advantages. They offer a refuge between opposing traffic streams so that cross traffic and pedestrians can traverse each stream as a separate maneuver. They also make space available for "left-turn" lanes; this clears the through lanes of turning vehicles and makes for smoother, safer operation and increased capacity.

Wide medians are preferred wherever space and cost considerations permit. For rural sections of freeways, 60 or 90 ft widths are common. *A Policy on Geometric Design* states, however, that 10 to 30 ft widths may be appropriate in suburban or mountainous situations. Medians up to several hundred feet have been provided in some instances, thereby completely isolating one roadway from the noise, confusion, and headlight glare of the other.

For rural and urban arterials, medians 60 ft wide or wider are recommended, since they allow the use of independent profiles and reduce crossover accidents. Medians 22 to 60 ft wide permit drivers to cross each roadway separately; widths of 14 to 22 provide protection at intersections for turning vehicles. Under very restricted conditions, curbed medians 4 to 6 ft wide may be employed as a means of separating opposing traffic, protecting pedestrians, and providing locations for traffic-control devices.

Wide medians generally are depressed below the level of the roadway, with the inside shoulder sloped toward the median for drainage. At times, however, medians, particularly narrow ones, may be curbed and crowned, with drainage across the traveled way. Even wide medians are at times graded up into a wide mound. Some of these alternatives are shown in *A Policy in Geometric Design*.

Very serious and spectacular accidents result when vehicle traveling at high speed cross the centerline or median and collide head on with or sideswipe those from the other direction. Concern over such occurrences has led to extensive research regarding the effectiveness of wide but traversable medians, and of various nontraversable barriers for positive separation.

With traversable medians, the width should be great enough to prevent most of the out-of-control vehicles from reaching the opposing traffic lanes. Fifty to 80 ft between lane edges has been suggested, but a specific value has not been stipulated. Cross slopes should not be greater than 6 to 1, with 10 to 1 preferred.

An alternative solution to wide medians combines a fairly wide median with some form of energy-absorbing device which will slow the vehicle without injury to its occupants. For example, it has been shown that dense, thick planting such as multiflora rose hedges, can be used safely as crash barriers.

For narrower medians, three means for reducing cross median accidents are being used. These are "deterring devices", "nontraversable energy-absorbing barriers", and "nontraversable rigid barriers".

Deterring medians incorporate such devices as two sets of double stripes painted on the existing pavement (called flush medians), raised diagonal bars, low curbing, and shallow ditches. They warn drivers and sometimes divert vehicles that are not too badly out of control back into the roadway. Sometimes they are designed to serve as refuges for left-turning or disabled vehicles. On the other hand, some fraction of vehicles cross into the opposing traffic lanes where the probability of a serious accident is high.

Nontraversable, energy-absorbing barriers are devices such as a chain-link fence about 3 ft high, supported on light steel posts and augmented by cables at bottom and midheight. When struck by vehicles, they prevent intrusion into opposing traffic but yield sufficiently to minimize the tendency to bounce vehicles back into the traveled way completely out of control. Sometimes vehicles become entangled in the fence and are brought to a halt. Although these barriers are effective, replacing damaged sections has been found to be costly as well as disruptive to traffic, since a lane must be closed while repairs are made.

Nontraversable rigid barriers are so designed that seldom does a vehicle get across this barrier into the opposing traffic lanes. On the other hand, it may be bounced back onto the roadway completely out of control and become involved in or be the cause of accidents there. Also, the upstream ends of some conformations of nonmountable curbs or railings provide a point of impact for vehicles, or turn them upside down. Until recently most of these barriers were very similar to a metal guard rail. The continuous horizontal element was of corrugated or box-beam cross section mounted on sturdy posts. Where medians were narrow, it was common to place barrier curbs along the inside shoulder and a guard-rail-type barrier with horizontal elements on both faces inside these curbs. Today, however, a high, nonmountable, sloped face concrete barrier called the New Jersey or GM. barrier is favored. This can be cast or extruded in place or precast in sections and set into position with a crane.

With narrow medians, headlight glare from vehicles traveling in one direction may be dangerous as well as annoying to motorists going the opposite way. Mounting a light barrier of lattice or expanded-metal on top of the vehicle barrier offers one solution to this problem.

In some instances median widths for major streets and expressways at grade are variable; wide at intersections and narrow between.

Openings through medians on freeways to permit U-turns, although they would be useful to a few motorists, create a serious accident hazard. *A Policy on Geometric Design* recommends against them, except in rural areas where the spacing between interchanges is more than 5 miles. They may also be needed as emergency crossovers on elevated or depressed urban freeways carrying heavy volumes of traffic. In either instance, use, unless under police direction, is to be restricted to emergency vehicles.

Words and Expressions

glare   *n.* 眩光
intersection   *n.* 平面交叉口
pedestrian   *n.* 行人
consideration   *n.* 考虑;因素;报酬
halt   *n.* 招呼站;停止
upstream   *n.* 上游;上行
box-beam   *n.* 箱形梁
crossover   *n.* 立体交叉,跨线桥
traffic stream   交通流;车流
multiflora rose   野蔷薇

UNIT 8

Text A Design of the Cross Section (Ⅰ)

Cross sections of typical highways of modern design are shown in Fig. 1. Dimensions for each element are based on careful analysis of the volume, character, and speed of traffic and of the characteristics of motor vehicles and their operators.

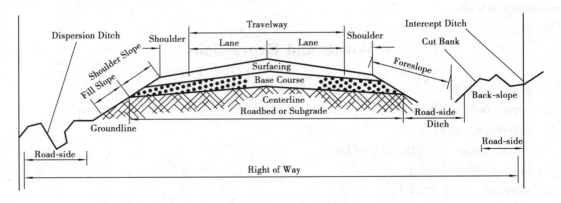

Fig. 1 Cross section of highway

Travelway In meeting oncoming vehicles or passing slower ones, the position selected by a driver depends primarily on the paved or surfaced width of the highway. Originally this surfaced width was only 15 ft, which was ample for horse-drawn vehicles. With the increase in motor-vehicle traffic and vehicle speeds, 24 ft widths of pavement are regarded as necessary for freeways and rural highways carrying high traffic volumes.

Shoulder The shoulder is that portion of the roadway between the outer edge of the traffic lane and the inside edge of the ditch, gutter, curb, or slope. Divided highways also may have an inside shoulder between the inside lane and the median. Shoulders provide a place for vehicles to park for changing tires, when otherwise disabled, or when stopped for any other reason[①]. If designs omit shoulders, or if they are narrow, roadway capacity decreases and accident opportunity may increase. The widths of 24 ft (two lanes at 12 ft each) are now preferred for high-type facilities.

Base course The portion of the materials above the subgrade that supports the weight of traffic. On earth surfaced roads there is no base course.

Surfacing The top layer of the travelway and shoulders; provides materials for maintenance blading and protects the underlying materials from traffic. No strength is assumed for the surfacing, although it has some when it is compacted.

Roadside ditch The ditch constructed at the bottom of a backslope for the purpose of collecting surface runoff water.

Intercept ditch A ditch located above the cutbank to collect runoff water and divert it from cutbanks that will erode.

Dispersion ditch A ditch located below the fill slope to receive collected water from drainage structures for returning to soils below the road.

Cross slope Except where superelevation of curves directs all water toward the inside, slopes usually fall in both directions from the center line of two-lane highways. For high-type pavements, this cross slope (or crown) is often 1/8 in. per foot to 1/4 in. per foot. For cheaper pavements constructed to less exacting standards the cross slope is greater. On paved shoulders cross slopes are usually greater, in the range of 3/8 in. to 1/2 in. per foot[2]. For gravel and turf even greater slopes are needed for satisfactory drainage.

Side slopes Earth fills of usual height stand safely with side slopes of 1.5 to 1. The side slopes of cuts through ordinary undisturbed earth remain in place with slopes of 1 to 1. Rock cuts as steep as 0.5 to 1 and sometimes 0.25 to 1 are stable. In the past, these slopes were standard for many highway agencies because they involved a minimum of earthwork. In recent years, however, side slopes generally have been flattened to provide for safer operation and decreased maintenance, for steep side slopes on fills create a serious accident hazard. If one wheel of a vehicle goes over the edge, the driver losses control. Overturn may result. With flat slopes the car can often be directed back onto the road or continue safely down the slope. Steep slopes on gutter ditches create similar accident hazards.

Steep slopes erode badly, thus creating serious maintenance problems. Furthermore, it is difficult to grow plants or grasses on them to aid in erosion control. Thus the saving in original excavation and embankment costs may be more than offset by increased maintenance through years; and in addition the slopes will be unsightly.

AASHTO standards now demand flat slopes on the roadway side of gutter ditches and at the top of the fill slopes. Standards for the Interstate System recommend that side slopes be no steeper than 4 to 1 and never steeper than 2 to 1 except in solid rock or other special soils[3].

Words and Expressions

cross section 横断面,横截面;断路面

typical   *a.* 典型的
dimension   *n.* 尺度,大小;度,维
character   *n.* 性质,特征,性格
characteristic   *n.& a.* 特点,特征,特色;特有的,表特征的
originally   *ad.* 本来,原来;独创性地
ample   *a.* 充分的,足够的
freeway   *n.* 高速公路
travelway   *n.* 行车道
curb   *v.* 路缘石,石;道牙;井栏
park   *v.* 停放,停车
median   *a.* 中间的,中央的; *n.* 中央分隔带,中间分车带
subgrade   *n.* 路基;路基面;地基
surfacing   *n.* 路面,面层;路面铺设,路面整修
blading   *n.* (用地平机)平路,刮路,整形,整平
compact   *v.* 压实,夯实
back slope   *n.* 后坡,内坡
runoff   *n.* 流出;径流;泄水
intercept   *n.* 截距;截线
divert   *v.* 转移,使转向
cutbank   *n.* 挖坡,边坡
erode   *v.* 冲蚀,侵蚀
dispersion   *n.* 分散,分散体,分散作用
superelevation   *n.* 超高
curve   *n.* (道路)线形
crown   *n.* 路拱,拱度
exacting   *a.* 严格的,苛刻的
gravel   *n.* 砾石,卵石
turf   *n.* 泥炭;草皮;草根土
undisturbed   *a.* 未搅动的;原状的,原来的
flatten   *v.* 压扁;整平
offset   *v.* 抵消,补偿
unsightly   *a.* 难看的,不堪入目的
earth fill   填土
in place   处于适当的位置;恰当的,合适的

Notes

①when 引导的是省略的状语从句,其逻辑主语是 vehicles,从句的完整形式是 when they are otherwise disabled, or when they are stopped for any other reason.本句可译为:车辆需要换胎,

或别的方面出了毛病,或因故受阻时,路肩可用作停车的地方。

②in the range of… to…意为 ranging from… to…或 ranging between… and…,"在……的范围内变化"。In.是 inch 的省略形式。本句可译为:在铺砌的路肩上,横坡的坡度通常比较大,其变化幅度可为 3/8~1/2 英寸。

③the Interstate System 指美国州际道路系统。Recommend 后面的宾语从句用虚拟语气第一型,即动词原形。本句可译为:按照美国州际道路系统的标准,边坡的坡度仅为 4∶1,除非在坚石或特殊土质的条件下,绝对不能超过 2∶1。

Text B Design of the Cross Section (Ⅱ)

The cross section of a typical highway in modern design is shown in Fig. 1.

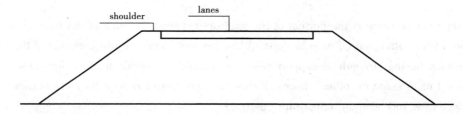

Fig. 1 Cross section

It is common practice in designing new highways to adopt a given cross section and employ it from end to end of the improvement. Seldom is this approach challenged on high-volume facilities. However, for low-volume facilities or for reconstructing old highways, it may be appropriate to modify features such as shoulder width in rough country or on longer bridges to reduce costs.

Lane Widths

In meeting oncoming vehicles or passing slower ones, the position selected by a driver depends primarily on the paved or surfaced width of the highway. Originally this surfaced width was only 15-ft which was ample or horse-drawn vehicle. With the increase in motor-vehicle traffic the width increased first to 16-ft, then to 18-ft. later two 10-ft lanes became a standard width for first-class paved highways. Today, 12-ft lanes are standard for freeways and other major traffic arterials although 14-ft widths have been recommended. For two-lane rural highways, a 24-ft-wide surface is required for clearance between commercial vehicles and is recommended for main highways. For collectors, surfacing widths of 20-ft are considered adequate only for low volumes and few for 20 mph design speed and ADT(Average Daily Traffic) less than 50, but range up to 22-ft. For urban streets, minimum design lane width is 12-ft; 11, 10, or even 9-ft are permitted where space is limited.

Recent research has brought another argument for wide lanes when there are frequent meetings or overtakings between passenger cars and large trucks, this is that, with strong cross winds, air disturbances can cause vehicles to swerve substantially within or even out of their lanes.

Number of Lanes

In almost all situations, the number of lanes in a new segment of highway is set by bringing together estimates of traffic for the design year and of highway, street, or lane capacity at the desired level of service. Four lanes in one direction in a single roadway has been the accepted maximum. However, AASHTO policy recognizes dual-divided roadway 16 lanes wide consisting of 4 lanes in each direction for an inner freeway with 4 more freeway lanes in each direction on the outside. In some instances reversible traffic flows between morning and night. Also exclusive bus lanes are at times provided. For mountainous areas the need for and location of climbing lanes for slow-moving vehicles can be explored on the basis of data. Changes in the number of lanes should not be made at interchanges or intersections.

Shoulders

The shoulder or verge is the portion of the roadway between the edge of the traffic lane and the edge of the ditch, gutter, curb, or side slope. AASHTO sets its usable width as that of the pavement or other surface having strength to support vehicles, shoulders provide a place for vehicles to stop when disabled or to stand for other reasons. If designs omit shoulders or if they are narrow, roadway capacity decreases and accident opportunity increases.

Shoulders on rural highways were originally 2, 3, or 4 ft wide and usually unpaved. Sometimes they were surfaced with gravel or similar material to provide hard standing at all times, but often they were earth and unusable in wet weather. Today, shoulders on major highways are usually paved. AASHTO recommends that when lane and shoulder are of bituminous construction, they be differentiated by color or texture. In the east, south, or mid-west of America where rainfall is sufficient and frequent enough to support grass, turf shoulders so constructed as to provide firm support for vehicles are sometimes used. American practice calls for shoulders continuous along the full length of the roadway on almost all roads. In certain European countries occasional turnouts (called laybys in Great Britain) may be used instead on all but major roads.

One argument for wide, continuous shoulders is that they add structural strength to the pavement. Others are that outside shoulders increase horizontal sight distance on curve and provide for snow storage during and after storms. Finally, they may reduce accident potential when vehicles stop for emergencies or other reasons.

Wide shoulders provide vehicles a place to stop off or temporary off the traveled lanes. Of these stops, those for emergencies are primarily related to traffic volume and flow conditions. But leisure stop varies with other factors such as trip length, rural versus urban, scenic versus nonscenic, and whether or not other places to stop are available. In any event, the argument for wide shoulders on all roads is inconclusive. For example, data from one study have indicated that on a mile of rural road carrying 400 vehicles per day, the chance of an accident involving parking maneuvers or standing or slow-moving vehicles is less than 1 in 100 per year. Again, it can be argued that, at least at lower traffic volumes, shoulders on long bridges or elevated structures are not justified on either accident-reduction or economic grounds.

It is now common practice to paint a continuous narrow white stripe to mark the line between roadway and shoulder as a guide for drivers during adverse weather and poor visibility conditions. The 1978 manual on uniform traffic control devices states that, "Edge lines shall be provided on all interstate highways and may be used on other classes of roads." The evidence is that when such a stripe is present, drivers tend to stay in the traffic lane and fewer of them infringe on the shoulder. Experience in Kansas on 450 m of highways has shown a 14% reduction in accidents and 25% reduction in deaths with edge lines. However, the evidence to date is not conclusive. Those opposed to striping argue that on two-lane highways, keeping vehicles in the traffic lane positions makes them closer to opposing traffic and makes head on collisions more likely.

For all freeways, *A Policy on Geometric Design* recommends that the outside shoulder be paved at least 10 ft wide, with 12 ft called for if truck volume is more than 250 in the design hour. Recommended width for the left (median) shoulder should be 10 ft wide, or 12 ft if truck volume in the design hour exceeds 250.

For rural arterials at ADTs less than 400, usable shoulder width is set at 4 ft minimum, with 8 ft suggested. It ranges to 8 ft minimum and 12 ft suggested when design hour volume exceeds 400. For urban arterials, where possible, similar shoulder widths without curbs, unless needed for drainage, are proposed. However, it is recognized that in many instances all available space is required for traffic so that shoulders must be omitted. The width of median shoulders on four-lane divided arterials is set at 3 ft minimum; for six or more lanes, 8 to 10 ft widths are recommended.

For rural collectors, a graded shoulder 2 ft wide for SDTs less than 400 to 8 ft at ADTs over 2000 is called for. In this case, width is defined as extending from the edge of the surfacing to the point where shoulder slope intersects side slope. Urban collectors usually do not have shoulders; rather, parking lanes 8 ft, or preferably 10 ft, wide and gutters are prescribed.

For local rural roads, a graded shoulder extending outside the surfacing to an intersection with the side slope is stipulated. Its width is 2 ft for ADTs less than 400, 4 ft for higher volumes.

Words and Expressions

shoulder   *n.* 路肩
oncoming   *a.* 即将来临的;接近的; *n.* 接近;来临
ample   *a.* 充足的;广大的;丰富的
overtake   *v.* 追上;超过
ditch   *n.* 排水沟
gutter   *n.* 排水沟
curb   *n.* 路缘石
stipulate   *v.* 讲定;确定;保证

UNIT 9

Text A Design of Intersections at Grade

Except for freeways, all highways have intersections at grade, so that the intersection area is a part of every connecting road or street. In this area must occur all crossing and turning movements.

The unchannelized intersections, as shown in fig.1(a) (c), are cheapest and least elaborate. With them, the intersecting roadways have been joined by circular arcs in order to provide pavement under vehicles turning to the right. For right-angle intersections of roads or streets carrying little traffic, no further treatment except signs may be deemed necessary, with the possible exception of curbings to keep vehicles on the pavement or to channel surface drainage. Y intersections or other conformations where vehicles meet at unfavorable angels may demand channelization, flared designs involve (1) widening the entering traffic lanes to permit deceleration clear of through traffic, and (2) widening the leaving lanes to provide for acceleration and merging. The channelized designs shown in fig.1(b), (d) are intended to direct approaching drivers to the particular paths by employing the principles of channelization enumerated above.

Fig. 1 shows several schematic layouts for situations where cross streets intersect divided expressways with frontage roads. If the frontage roads were omitted in Fig. 1(d), it would also be typical. Note the acceleration and deceleration lanes to clear the through lanes of both right-turn and left-turn vehicles, channelizing islands, and pavement markings. A possible bus routing is suggested on Fig. 1(d).

A careful traffic count and estimate of future changes for all movements, including turns, must precede the design of important intersections. Only in this way can those movements that are heavy be favored. This coupled with knowledge of lane capacities leads to decisions regarding the number of lanes to be supplied. The speeds at which vehicles approach and move through the intersection and vehicle type govern many dimensions, particularly minimum sight distances in various directions, the radii of curves, and the lengths of the various turning and storage lanes. Likewise,

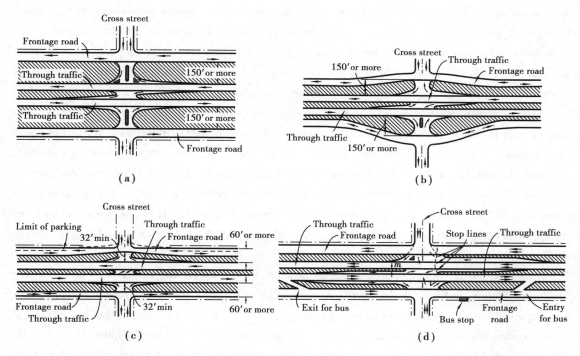

Fig. 1 Expressway intersections with frontage roads

(a) two-way frontage roads, wide outer separation; (b) two-way frontage roads, bulbed separation;

(c) two-way frontage roads, minimum design; (d) one-way frontage roads, minimum design

a decision regarding the need for traffic signals currently or in the future affects certain features of the design.

In laying out intersections, characteristics of driver and vehicle and the possibility of accidents and their frequency and severity must be always kept in mind. As stated earlier, drivers should be confronted with one decision at a time. Furthermore, they should be guided into the proper channels for their intended routes and prevented from doing wrong or unpredictable things. It is important that spacious areas which permit "open-field running" be eliminated by the provision of directional islands that leave little choice of route. Pavement and islands at very small angles. It is preferable to use single large islands rather than several small ones, as single large ones are less confusing to drivers.

Adequate sight distances for vehicles entering intersections at anticipated speed are required for safe operation. These encompass, first, seeing the roadway as the driver approaches, passes through, and travels beyond the intersection and, second, a clear view of vehicles that may be approaching on other intersection legs. For the latter situations, graphical *minimum sight triangles* drawn on a plan view of the intersection can be employed to determine available or needed sight distance. These would differ between situations where vehicles proceed through the intersection without stopping and where entry on some or all the legs is controlled by stop or yield signs or traffic signals. Detailed procedures for making such analyses are offered in *A Policy on Geometric Design*.

Building, signs, plantings, or other developments on adjoining private property which impair sight distance are often a problem. These cannot be removed by agency personnel, as that would infringe on private rights. In many jurisdictions, ordinances limiting the height of plantings or stipulating setbacks are enacted, but if owners refuse to comply, court action may be necessary. Violators also may be subjects to suits for negligence by injured motorists if they can prove the obstruction was a major contributor to an accident.

Most important intersections must accommodate large trucks and the radii of all curves made long enough for them. As an illustration, for the inside edge of 90° turning roadways at low-speed intersections, AASHTO recommends, as a minimum, three centered compound curves with radii successively 180, 65, and 180 ft, with the radius point of the 65-ft curve 71 in from the tangent. Because vehicles do not track and their fronts overhang, minimum lane width near the center of the curve is 20 ft. in addition, a raised island having an area of about 125 ft is interposed between the turning roadway and the through lane that it is joining. *A Policy on Geometric Design* gives details such as these for typical situations.

In settled areas, street and intersection designs must consider the needs of pedestrians and bicyclists. Walking rates for pedestrians range from 2.5 to 6.0 ft/s with older people moving at the lower speeds. For design purposes in timing traffic signals, 4.0 ft/s is a common value but 3.0 ft/s is recommended where a substantial number of older people are involved. Widths of crosswalks commonly are 4.0 ft in residential areas and 6.0 ft in commercial areas. If bicyclists traveling in two directions are to be accommodated, a minimum width of 6.5 ft is recommended. With large pedestrian volumes, greater crosswalk widths than those listed may be needed.

Sometimes a study of traffic-flow data indicates that relatively few vehicles perform a particular turning movements at a given intersection. For example, this could be expected where the streets intersect at oblique angles. If this movement complicates the design or increase congestion, it should be eliminated completely and provision made for it in some other way. To illustrate, vehicles desiring to make left turns often are directed to the right completely around an adjacent block; then they pass through the intersection with normal cross traffic.

Where left-turn movements between certain legs of an intersection are extremely heavy, two lanes rather than one may be provided in or adjacent to the median on the approach leg. In such cases, traffic signal timing is such that this path is free of opposing traffic.

Certain design features complicated intersections often require testing by actual use. In such case it is common to place temporary barricades or channelizing islands of sandbags which can be easily shifted around. Sand sprinkled over the roadways indicates vehicle paths. After the design has been proved, the permanent installation is made. The preceding paragraphs have merely suggested the complexities of proper intersection design. For added detail, including dimensional requirements to fir almost all situations and for examples of good designs, the reader is referred to *A Policy on Geometric Design*. It gives numerous examples of special intersection conformations to fit unusual situations.

Words and Expressions

converge   *v.* 集中;汇合;收敛
unpredictable   *a.* 不可预测的; *n.* 不可预测的事物
enact   *v.* 扮演;规定;颁布
infringe   *v.* 侵犯;违背;侵害
jurisdiction   *n.* 司法权;审判权;管区;权限
ordinance   *n.* 法令;宗教仪式
setback   *n.* 挫折;逆流;倒退
oblique   *a.* 斜的
sandbag   *n.* 沙袋
frontage road   前沿道路(沿临街房屋前面的地方道路或辅助道路)

Text B   Interchanges

An interchange is a grade separation in which vehicles moving in one direction of flow may transfer by the use of connecting roadways. These connecting roadways at interchanges are called ramps.

Fig. 1 shows typical layouts of interchanges at three-legged junctions. Fig. 1(a) shows a T or trumpet interchange pattern suitable for three-way intersections. Note that traffic from lower to upper left must traverse a 270° turn but that all other turning movements are accomplished with curvatures not much greater than 90°. Fig. 1(b) portrays a layout for a Y intersection. Here only one grade separation is required to eliminate all crossings at grade. It should be noted, however, that provision for vehicles traveling from lower to upper left is circuitous.

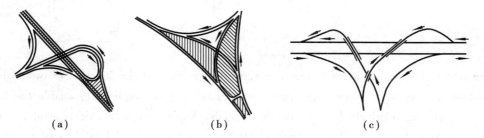

(a)   (b)   (c)

Fig. 1   T- and Y-interchanges
(a) T- or Trumpet interchanges; (b) Y-interchanges; (c) Semi-direct connection for T-junction

Also it requires two weaving motions and leaving one roadway and entering another on the left. It can be argued that this movement should be prohibited entirely. Fig. 1(c) indicates an intersection where all turning movements are facilitated in this way.

The simplest type of four-way interchange is the Diamond(Fig. 2) consisting of a single bridge and four one-way ramps. It is particularly adapted to situations where a freeway crosses a non-freeway

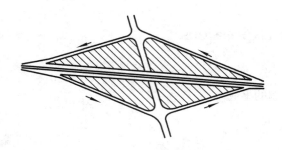

Fig. 2   Diamond interchange

arterial. Flow on the freeway is uninterrupted, except for problems that may develop at points where ramp traffic enters or leaves. But traffic patterns on the arterial are complex, since the roadway must carry two through movements and accommodate four left turns, two of which must use the inside lanes of the arterial or separate turning lanes. When volumes are large, traffic signals generally are required.

Probably the most common interchanges where freeways intersect arterials is the cloverleaf [Fig. 3(a)]. It is often regarded by motorists as the ultimate answer to intersection problems. It has the great advantage of being very uncomplicated to use. With it, the intersecting arterials are separated, and in addition all eight turning movements are accomplished free of intersections where vehicle paths must cross. Turning vehicles peel off the right side of the roadway on which they enter the interchange and blend from the right into the roadway being entered[①]. Nevertheless there are a number of features about this type of interchange which limit its usefulness.

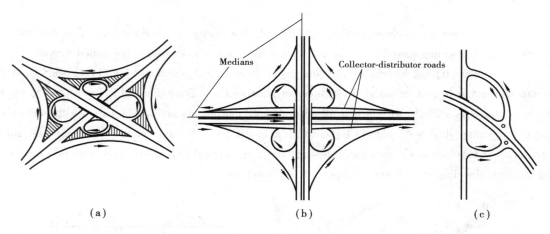

Fig. 3   Cloverleaf

The first and most important is that if a Cloverleaf is used at the junction of two high-speed heavy-volume highways an excessively large area of land may be required to enable the loops to handle the traffic at relatively low speeds differentials. Experience with this type of interchange indicates that loop design speeds in excess of 50 km/h and loop radii less than 61 m are rarely justified[②]. At greater speeds, extra time is required to travel the necessarily longer loop distances. Shorter radii can be dangerous and yet not reduce the land area requirement by very much, i.e. the length required for a 61 m radius is generally about the minimum required to comfortably secure the necessary difference in elevation between the two roadways[③].

A second undesirable feature of the Cloverleaf is that vehicles wishing to make right-hand turning movements must negotiate a 270 degree semi-direct turn. Not only can this represent a relatively difficult design problem but, in addition, as vehicles leave a particular loop upon entering the main highway, it is necessary for them to weave their way through other vehicles attempting to enter the adjacent exit loop[④]. Therefore, when traffic volumes are heavy, the area required for the Cloverleaf may have to be considerably enlarged unless ancillary weaving lanes are provided between the loops.

Fig. 3(b) diagrams an alternative Cloverleaf design with collector-distributor roads. These can be applied to one or both through roadways if the costs of added land, paving, and structures can be justified. With this design, weaving and merging movements are separated. Also, the collector-distributor road provides an opportunity for speed adjustment clear of the freeway.

The partial-cloverleaf [see Fig. 3(c)] offers connections by merging to the major freeway but calls for left turns through opposing traffic on the minor artery. It can be developed in many forms with the loops in different quadrants to fit topography and traffic patterns.

Where freeways meet freeways and all traffic movements are heavy, interchanges often are designed with directional left turns in all four quadrants. The outstanding design characteristic of this type of interchange is the use of a high design speed throughout, with curved ramps and roadways of large radius. The land requirements for a directional interchange are therefore very large. In cases where volumes for certain turning movements are small, design speeds for these movements are reduced and the turnoff is effected within a loop. Fig. 4(a) shows examples of directional interchanges.

Recently, experience with directional interchanges has revealed operational problems, associated with left-hand entrance and exit ramps. Most drivers expect to exit freeways to the right and to enter from the right. When those expectations are violated, confusion, erratic maneuvers and accidents sometime result.

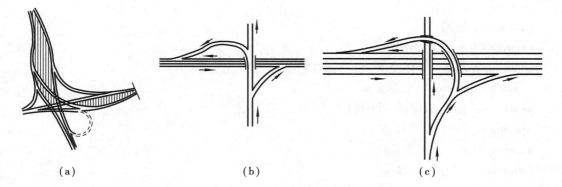

(a)　　　　　　　　　　　　(b)　　　　　　　　　　　　(c)

Fig. 4　Directional interchanges

(a) Directional interchange; (b) One-exit interchange; (c) Two-exit interchange

Consider the alternative designs shown in Fig. 4(b), (c). Design (b) is a directional interchange in which a driver wishing to go left. A northbound driver must make two decisions as he approaches the first exit: (1) that he is going to leave the freeway at this exit, and (2) whether his destination is to the right (east) or left (west). Confronted with such interchange configuration, a driver may be in the right-hand lane, decide that he must go west to reach his destination, and be faced with a need to cross several lanes of high-speed traffic in order to exit to the left.

Design (c) is a preferred single exit configuration on which a northbound driver makes decisions one at a time. The driver exits to the right and then decides whether to proceed east or west while on the lower-speed, less congested exit ramp[⑤].

Words and Expressions

interchange   *n.* 互通式立交;道路立体枢纽
T- and Y-interchange   T 形和 Y 形互通立交
diamond interchange   菱形立交
directional interchange   定向立交
trumpet interchange   喇叭式立交
ramp   *n.* 匝道;坡道;斜道
ramp traffic   匝道交通
diamond ramp   菱形匝道
curved ramp   弯曲匝道
entrance ramp   驶入匝道
exit ramp   驶出匝道
layout   *n.* 布置;规划;设计
junction   *n.* 连接;汇合点,交叉处
three-legged junction   Y 形交叉口
traverse   *v.* 穿过,经过
curvature   *v.* 弯曲,曲率
portray   *v.* 描绘
circuitous   *a.* 迂回路线的
grade separation   立体交叉(公路,铁路)
at grade separation   平面交叉
weave   *v.* 编织,编排;使迂回前进;组合
weaving motion   交叉行进
weaving lanes   交叉道路
facilitate   *v.* 促进,使容易
accommodate   *vt.* 向……提供,供给;容纳
cloverleaf   *n.* 苜蓿叶
partial cloverleaf   半苜蓿叶
full cloverleaf   全苜蓿叶
peel   *v.* 剥落,脱落;削皮
peel off   离群,离队
loop   *n.* 环路,环线,环道
justify   *v.* 证明……是正当的
elevate   *v.* 提升,提高

elevation　　*n.* 提升;标高,高程;立视图
negotiate　　*n.* 通过,越过
adjacent　　*a.* 临近的,接近的
ancillary　　*a.* 辅助的,副的
merge　　*v.* 合并,并入;融合
quadrant　　*n.* 象限,圆周的四分之一,扇形板
turnoff　　*n.* 岔道
violate　　*v.* 违反;妨碍;干扰
erratic　　*a.* 移动的,不规律的
northbound　　*a.* 北行的
configuration　　*n.* 形状,外形;地形,线路
congest　　*v.* 拥挤,拥塞;充满
collector-distributor road　　集散道路
through roadway　　直通道路
except for　　除……之外
one at a time　　一次(做)一个(决定),分别

Notes

①Turning vehicles peel off the right side of the roadway on which they enter the interchange and blend from the right into the roadway being entered. 句中 peel off 意为"分离出来",blend from the right into 意为"从右侧汇入"。本句可译为:转弯车辆从主线的右侧分离,并从另一条主线的右侧汇入。

②in excess of:超过,例如:to spend in excess of one's income.开支多于收入。km/h 指"千米/小时"。本句可译为:使用这类立交的经验表明:环道设计车速超过 50 km 而环道半径小于 61 m 很少被认为是合理的。

③the length required for a 61 m radius is generally about the minimum required to comfortably secure the necessary difference in elevation between the two roadways.句中的 required for a 61 m radius 和 required to comfortably secure the necessary difference in elevation between the two roadways 是-en 分词短语分别作定语修饰 the length 和 the minimum;difference in elevation 指"标高差;高程差"。本句可译为:61 m 半径所需的长度通常是确保两条道路之间必要的标高差所需的最小值。

④not only… but also…,是最常见的搭配,但它有一些变体,如 also 可以省略,也可由 as well,in addition 之类代替。例如:I not only heard it… but saw it.我不但听见了,而且看到了。not only 置于句首时,主句需要倒装。本句中 as 引导的是时间状语从句,attempting 引导的分词短语作状语,修饰 weave their way…。本句可译为:这不仅会给设计带来相当大的问题,而且当车辆离开环道进入主线时,它们必须与其他试图进入相邻环道的车辆交织运行。

⑤while…是省略式状语从句,on 前省去了 he is。本句可译为:驾驶员从右边驶出主线,然后决定在速度低、车辆少的驶出匝道上是向东去还是向西去。

UNIT 10

Text A Portland Cement

Portland cement is credited to an Englishman named Joseph Aspdin, a mason, who obtained a patent for his product in 1824. He named it Portland cement because it produced a concrete that resembled a natural lime-stone quarried on the Isle of Portland in the English Channel.

Portland cement is produced in a plant where raw materials are heated in a rotary kiln. The high heat in the kiln causes a chemical reaction that converts the raw materials to clinker. The clinker is then pulverized to form the cement.

The four major chemical components that must be present in the raw materials in order to produce Portland cement are Lime, Iron, Silica, and Alumina. Sources for these raw materials are illustrated in Tab. 1. Note that some raw materials are available from the same source.

Tab. 1 Sources of major components in Portland cement

Lime(CaO)	Iron (Fe_2O_3)	Silica(SiO_2)	Alumina(Al_2O_3)
Calcite	Clay	Clay	Aluminum ore refuse
Limestone	Iron ore	Marl	Clay
Marl	Mill scale	Sand	Fly ash
Shale		Shale	Shale
Aragonite			
Also used: Anhydrite or Gypsum, $CaSO_4 \cdot 2H_2O$			

The raw materials can be combined in either a dry or a wet process. Here depict the steps involved in manufacturing Portland cement by the dry process.

Step 1: The raw material is first put through a crusher to break it down into 5-inch, or smaller, size.

Step 2: Raw materials are proportioned, put through another grinding, and then dry mixed. This provides the proper chemical composition for making the clinker.

Step 3: The raw materials are preheated, and then be put into the kiln where temperatures are between 2,600 to 3,000 degrees F and where they are chemically changed into cement clinker.

Step 4: As the clinker exits the kiln, it is air-cooled. The rate of cooling must be controlled as it can affect the properties of the cement.

Step 5: The clinker is cooled, pulverized, and then stored or shipped.

Note the addition of gypsum which is added to control setting time and helps in obtaining optimum strength.

After the final grinding, the cement is a fine powder whose particles are so small that they could pass through a sieve with 40,000 openings per square inch.

Cement Types

Different types of cement are manufactured to meet different physical requirements and for specific requirements and for specific purposes. There are five basic types of cement, as follows:

- Type I or I A (air entraining) is a general purpose cement that is used when the special properties of the other types are not required.
- Type II or II A (air entraining) is used where there is a need to protect against moderate sulfate attack. Naturally occurring sulfates such as sodium and magnesium are sometimes found in soils or dissolved in groundwater where concrete is to be placed.
- Type III or III A (air entraining) is a high-early-strength cement used where high strength concrete is needed in a short period, usually about one week. Uses would include situations where forms must be removed early, where a structure has to be put into service quickly, or in cold weather.
- Type IV is a low heat of hydration cement that is used in large structures such as dams where a heat rise is a critical factor. Excessive heat rise could cause cracking due to volume change.
- Type V is a high sulfate-resistant cement.

It is possible specify a white Portland cement; however, this is used mostly for architectural purposes or for color contrast, such as in curbs and median barriers.

Blended cements consist of Portland cement combined with some other ingredient. There are several types of blended cements depending upon what is added to replace the cement in the clinker. Type I S cements are ones that have had blast-furnace slag added. As much as 65% to 75% slag may be substituted for the cement. Type I P cements have had a Pozzolan added in place of a portion of the cement clinker.

A Pozzolan is a siliceous or siliceous and aluminous material, which in itself possesses little or no cementitious value but will react chemically with water and calcium hydroxide to form compounds possessing cementitious properties. The most common Pozzolan is Fly-Ash. Here again, intergrinding is the common manufacturing process and the Pozzolan accounts for between 15% and 40% of the weight of the cement.

Just briefly, some of the advantages of using a Pozzolan are: improved workability, economy, reduce alkali-aggregate reaction, increase sulfate resistance, and reduced heat generation, volume change and bleeding.

Cement Compounds

When the clinker is formed during the manufacturing process, four principal compounds are created. Tab. 2 represents a typical composition.

Tab. 2 Composition of cements

Type	Compound(%)				Wagner fineness(m^2/g)
	C_3S	C_2S	C_3A	C_4AF	
I	55	19	10	7	0.18
II	51	24	6	11	0.18
III	56	19	10	7	0.26
IV	28	49	4	12	0.19
V	38	43	4	9	
White	33	46	15	2	

- Tricalcium silicate ($3CaO \cdot SiO_2$ abbreviated C_3S) hardens rapidly and is responsible for initial set and early strength. Note that Type III cement (high early strength) is made up of 56 percent tricalcium silicate.
- Dicalcium silicate ($2CaO \cdot SiO_2$ or C_2S) hardens slowly, thus it is responsible for low heat of hydration and ultimate strength. Type IV cement (low heat of hydration) has the highest percentage of dicalcium silicate.
- Tricalcium aluminate ($3CaO \cdot Al_2O_3$ or C_3A) liberates large amounts of heat in the first days of hardening and contributes to early strength development. High amounts of this material, however, contribute adversely to the resistance to sulfate attack. Consequently, Type III cement (high early strength) has a higher percentage of C_3A than Type V cement (low sulfate).
- Tetracalcium Aluminoferrite ($4CaO \cdot Al_2O_3 \cdot Fe_2O_3$ or C_4AF) reduces the clinkering temperature, thus assisting in the manufacturing process. It acts like a flux in burning the clinker.

Words and Expressions

be credited to   归功于

mason   *n.* 泥瓦匠
quarry   *v.* 挖掘
kiln   *n.* 窑;炉
rotary kiln   回转窑
clinker   *n.* 炉渣;煤渣;熟料
pulverize   *v.* 粉碎;将……弄成粉末或尘埃
calcite   *n.* 方解石
limestone   *n.* 石灰石
marl   *n.* 泥灰岩
shale   *n.* 页岩
aragonite   *n.* 文石
clay   *n.* 黏土
iron ore   铁矿石
mill scale   轧屑
aluminum ore refuse   铝矿石垃圾
fly ash   粉煤灰
anhydrite   *n.* 硬石膏
gypsum   *n.* 石膏
depict   *v.* 描述
crusher   *n.* 破碎机
grind   *v.* 碾,磨
setting time   凝结时间
sieve   *n.* 筛子
openings   *n.* 孔
forms   *n.* 模板
cracking   *n.* 开裂
ingredient   *n.* 成分;(混合物的)组成部分
blast-furnace slag   高炉矿渣
pozzolan   *n.* 火山灰
cementitious   *a.* 胶凝的
accounts for   占……比例
workability   *n.* 工作性
alkali-aggregate reaction   碱集料反应
bleeding   *n.* 泌水
flux   *n.* 熔剂;助剂

Text B Cement Properties and Storage

Cement Properties

The properties expected in the Portland cement are carefully set out in the specifications of the various transportation agencies. Frequently this is done by reference to those prescribed by AASHTO. To meet these specifications, samples must pass a number of chemical and physical tests which can only be conducted in a well-equipped laboratory.

The chemical tests constitute, in effect, a chemical analysis to determine whether the various strength-giving compounds appear in proper quantity, and if there are excessive amounts of certain undesirable substances, while the physical tests constitute Fineness, Soundness, Consistency, Setting time, Compressive strength, Loss on ignition, Heat of hydration, Air content. Let's look at some of these in more detail.

Fineness

We have mentioned that the ground-up clinker is fine enough to pass through a mesh with 40,000 openings per square inch.

Fineness of the cement affects the rate of hydration. Greater fineness increases the surface available for hydration causing greater early strength and more rapid generation of heat. Type Ⅲ cement therefore has a high fineness. Because of the extremely small size of the particles, they do not lend themselves to analysis by means of sieving. Special methods have been developed to make approximations of the size distribution.

The measure of fineness is known as specific surface and is the summation of the surface area, in square centimeters, of all particles in one gram of cement. Cements with the fineness below 2,800 may produce concrete with poor workability and excessive bleeding.

Soundness

Soundness refers to the ability of hardened cement paste to retain its volume after setting. Expansion of the cement is caused by excessive amounts of lime or magnesia in the cement. The test for cement soundness is the Autoclave expansion test. One-inch by one-inch cross section molds used to form cement paste specimens 11 and 1/4 inches long. After curing for 24 hours, the specimens are measured for length and then placed into the autoclave. The specimens will be subjected to saturated steam at 295 PSI[①] for 3 hours. The specimens are removed, cooled and remeasured. The difference in length of the specimen before and after autoclave expansion and is measured to the nearest 0.01 percent.

Setting Time and Consistency

The cement should not set up too soon nor should it occur too late. Setting time tests are performed to determine if normal hydration is taking place in the first few hours. Demonstrated in Fig. 1 is the Vicat apparatus in which a needle is allowed to settle into cement paste. Time is recorded until the needle penetrates 25 mm. This is the initial setting time. Final set is when the needle stops sinking.

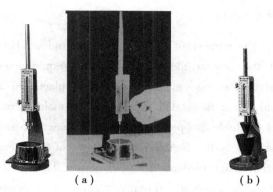

Fig. 1 Vicat apparatus for cement setting time tests and consistency determine

The Vicat apparatus is also used to determine the consistency of cement. Consistency means the cement's ability to flow. In this test, the needle is replaced with a 10 cm diameter plunger[②]. Normal consistency is defined as a 10 mm drop of the plunger in 30 seconds. The moisture content used to obtain this consistency is then used for mixing the paste in the setting time test.

Compressive Strength

ASTM C 150 and AASHTO M 85 specify certain minimum cement compressive strength requirements. Compressive strengths of cement mortars made using Ottawa sand are conducted. The test is performed on two inch mortar cubes produced in these molds as shown in Fig. 2.

For a Type I cement, 7-day strength of one inch mortar cubes should not be less than 2,800 PSI. Using that as a reference, Fig. 3 shows the relative compressive strength requirements for the various cements at various ages.

Fig. 2 Cubic mold for cement mortar

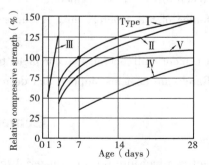

Fig. 3 Relative compressive strength requirements for Portland cements (ASTM C 150)

Loss on Ignition

The cement sample is heated to 1,000 degree C until it reaches a constant mass. The mass loss can then be calculated. A high loss of 3 percent or more is an indication of prehydration and carbonation. This may be caused by improper or prolonged storage.

Air Content

The purpose of the test is to determine whether or not the cement meets air-entrained requirements. To perform the test, a cement mortar is placed in a flow mold, on a flow table (Fig. 4). The mold is removed and the table dropped 1/2 inch, 10 times in 6 seconds. The flow is the increase in diameter of the specimen. Various water contents are tried until a flow of 80 to 95 percent is achieved. Through weighing a predetermined volume and knowing the percentage of moisture, the air content can be determined.

Fig. 4　Flow table

There are several other physical properties of cement should be mentioned.

False Set. This is a significant loss of plasticity shortly after mixing. This presents no problem if the concrete is remixed before it is placed. False set is an erratic condition that can be blamed on the character of the gypsum used.

Specific gravity of cement is about 3.15 and is used in the mix design calculations.

Weight of cement in the United States is usually measured by the bag at 94 pounds. Density of cement is not usually considered since it can be fluffed up quite easily. For this reason, cement is weighed for each batch of concrete.

Hot or green cement is sometimes encountered. Sometimes during peak periods cement plants will have difficulty keeping up with demand. Consequently, cement may be coming directly from production to the concrete plant. Also, cement may become hot due to silos being located in direct sunlight. When this happens, cement at elevated temperatures could be used in concrete. This could cause stiffening of the concrete mix. An upper limit of 150 to 180 degrees F is recommended for cement.

Each cement mill operates a complete testing laboratory and maintains close control on its product. Rarely are cements shipped that do not meet specifications.

Cement Storage

Cement is a moisture sensitive material that will retain its quality indefinitely if it is kept dry. When storing bagged cement, a shed or warehouse is preferred. Cracks and openings in store houses should be as low as possible, with bags stacked on pallets and not against an outside wall. When storing bagged cement outdoors, they should be stacked on pallets and covered all over with a waterproof covering (Fig. 5); they should not be stored where they will get wet from ponded or runoff rain water.

UNIT 10

Fig. 5 Storage of bulk cement

Most cement that arrives on a construction project will be by bulk. Storage of bulk cement should be in a watertight bin or silo. Separate compartments or bins should be provided for each type of cement. Transportation should be in vehicles with watertight and properly sealed lids. Cement conveyance systems, such as screw conveyors or air slides, should provide for constant flow and precise cutoff.

The result of improper storage and handling is lumpy cement. Loss on ignition or strength tests should be performed on cements that have been stored for long periods of time.

Words and Expressions

set out   （清晰而有条理地）陈述,阐述,说明
specifications   *n.* 规范
prescribe   *v.* 规定
fineness   *n.* 细度
soundness   *n.* 安定性
consistency   *n.* 稠度
setting time   *n.* 凝结时间
loss on ignition   *n.* 烧失量
heat of hydration   *n.* 水化热
air content   *n.* 含气量
mesh   *n.* 网孔
lend themselves to   适合（自己）
specific surface   *n.* 比表面积
summation   *n.* 总和
lime   *n.* 石灰
magnesia   *n.* 氧化镁
autoclave expansion test   *n.* 压蒸法;高压釜膨胀试验
cross section   *n.* 横截面
cement paste   *n.* 水泥浆

curing   *n.* 养护
specimen   *n.* 试件
saturated steam   *n.* 饱和蒸汽
Vicat apparatus   *n.* 维卡仪(水泥稠度及凝结时间测定仪)
penetrate   *v.* 刺入
diameter   *n.* 直径
specify   *v.* 规定
conduct   *v.* 实施
cube   *n.* 立方体
carbonation   *n.* 碳化
prolonged   *a.* 长期的
air-entrained   *a.* 加气的
percentage of moisture   *n.* 含水率
erratic   *a.* 不稳定的
specific gravity   *n.* 比重
batch   *n.* 一批;一批生产的量
silo   *n.* 筒仓;储藏库
stiffening   *v.* 硬化
mill   *n.* 工厂
ship   *v.* 运送;运输;船运
shed   *n.* 棚库
warehouse   *n.* 仓库
stack   *v.* 堆放
pallet   *n.* 托盘;平台;运货板;草垫子
by bulk   散装
watertight   *a.* 水密的;不漏水的;防渗的
bin   *n.* 仓
screw conveyors   *n.* 螺旋输送机
air slides   *n.* 空气滑道
cutoff   *n.* 终止
lumpy cement   块状水泥

Notes

①PSI：pounds per square inch
②plunger：a part of a piece of equipment that can be pushed down 活塞、柱塞

UNIT 11

Text A　Plain Concrete, Reinforced Concrete, and Prestressed Concrete

Plain Concrete

Concrete is a stone-like material obtained by permitting a carefully proportioned mixture of cement, sand and gravel or other aggregate, and water to harden in forms of the shape and dimensions of the desired structure. The bulk of the material consists of fine and coarse aggregate. Cement and water interact chemically to bind the aggregate particles into a solid mass. Additional water, over and above that needed for this chemical reaction, is necessary to give the mixture the workability that enables it to fill the forms prior to hardening. Concretes in a wide range of strength properties can be obtained by appropriate adjustment of the proportions of the constituent materials. Special cements (such as high-early-strength cements), special aggregates (such as various lightweight or heavyweight aggregates), and special curing methods (such as steam-curing) permit an even wider variety of properties to be obtained.

These properties depend to a very substantial degree on the proportions of the mix, on the thoroughness with which the various constituents are intermixed, and on the conditions of humidity and temperature in which the mix is maintained from the moment it is placed in the forms until it is fully hardened. The process of controlling these conditions is known as curing. To protect against the unintentional production of substandard concrete, a high degree of skillful control and supervision is necessary throughout the process, from the proportioning by weight of the individual components, through mixing and placing until the completion of curing.

The factors which make concrete a universal building material are so pronounced that it has been used for thousands of years, probably beginning in Egyptian antiquity. The facility with which, while plastic, it can be deposited and made to fill forms or molds of almost any practical shape is one of these factors. Its high fire and weather resistance are evident advantages. Most of the constituent materials, with the possible exception of cement, are usually available at low cost locally or at small distances from the construction site. Its compressive strength, like that of natural stones, is high, which makes it suitable for the members primarily subject to compression, such as columns and arches. On the other hand, again as in natural stones, it is a relatively brittle material whose tensile strength is small compared with its compressive strength. Hence, plain concrete is used only for footings and concrete slabs laid on the ground, and for massive structures such as retaining walls; even then reinforcement is frequently employed.

Reinforced Concrete

To offset the limitation of plain concrete, it has been found possible, in the second half of the nineteenth century, to use steel with high tensile strength to reinforce concrete, chiefly in those places where its small tensile strength would limit the carrying of the member. The reinforcement, usually round steel rods with appropriate surface deformations to provide interlocking, is placed in the forms in advance of the concrete. When completely surrounded by the hardened concrete mass, it forms an integral part of the member. The resulting combination of two materials, known as reinforced concrete, combines many of the advantages of each: the relatively low cost, good weather and fire resistance, good compressive strength, and excellent formability of concrete and the high tensile strength and much greater ductility and toughness of steel. It is this combination which allows the almost unlimited range of uses and possibilities of reinforced concrete in the construction of buildings, bridges, dams, tanks, reservoirs, and a host of other structure.

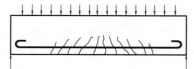

Fig. 1 Reinforced concrete cracks under working loads

The steel bars in concrete take the tensile component of the bending moment. But they do not prevent the concrete from cracking (Fig. 1).

The presence of fine cracks in reinforced concrete is inevitable. The stress in the lowest-grade reinforcing steel under the normal working loads is of the order of 140 MPa. Taking the modulus of elasticity of steel as 200 GPa, this amount to an elastic strain of 7×10^{-4}, which is more than the ultimate tensile strain of concrete. Cracks are thus produced in the concrete by the mere process of the reinforcing steel being stressed under the normal working loads. It is perhaps fortunate that this was not understood when reinforced concrete was first employed more than a century ago, otherwise building authorities with a reasonable concern for public safety would probably have forbidden its use.

Because if the cracks are kept very small and are bridged by tension steel they have no adverse effect on the safety or durability of the structure, the safety of reinforced concrete structures depends on the width of the cracks being kept below a permissible minimum and this has become a more serious problem in recent years because the use of higher steel stresses also increases the strain of the concrete. Cracks would not only be unsightly but would expose the steel bars to corrosion by moisture and other chemical action. Thus crack control is a more serious matter in reinforced concrete design as compared with, say, twenty years ago.

Prestressed Concrete

Methods of inducing compression in concrete member before it is loaded is known as prestressing. The construction which uses steels and concrete of very high strength in combination is known as prestressed concrete(Fig. 2).

The steels, mostly in the shape of wires of or strands but sometimes as bars, is embedded in the concrete under high tension which is held in equilibrium by compression stresses in the surrounding concrete after hardening. Because of this compression, the concrete in a flexural member will crack on the tension side at much larger load than when not so pre-compressed. This reduces radically both the deflections and the tensile cracks at service loads in such structures and thereby enables these high-strength materials to be used effectively.

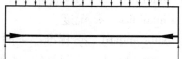

Fig. 2 Prestressed concrete cracks only under an overload

Prestressed concrete is particularly suited to prefabrication on a mass-production basis. Its introduction has extended, to a very significant degree, the range of structural uses of the combination of these two materials.

Prestressed is being successfully used in the manufacture of concrete sleepers. Ordinary reinforced concrete sleepers quickly fail in service due to the rapid opening and closing of cracks when loads are applied and removed in rapid succession. It has been found that some prestressing sleepers do not crack, and disintegration due to the opening and closing of cracks is therefore avoided.

Hollow beams for bridge and concrete piles constructed of precast blocks are examples of the use of prestressed members. It is not possible to predict the length required for a pile, and it is a great convenience to be able to extend a pile during driving by adding blocks; the handing of long pile is avoided and, provided the lateral support of the ground is adequate, the prestressing cables can be removed upon completion of driving and used again.

Another advantage of prestressed concrete is that the concrete and the steel are severely tested during the prestressing operation, and a lower factor of safety is justified. The permissible working stress in the concrete is generally one-third of the compressive strength, thus allowing a margin to cover the risk of poor concrete occurring at a critical section. The risk is reduced by prestressing, because the stress induced in the concrete during the prestressing operation may be 50% to 75% of its compressive strength.

Words and Expressions

    plain concrete   素混凝土
    interact   *v.* 相互作用

proportioned　*a.* 成比例的

coarse aggregate　粗集料（粒径粗于 5 mm）

fine aggregate　细集料

forms　*n.* （灌混凝土用）模板，模型，模槽

constituent　*a.* & *n.* 组成的；构成（部分），成分，要素

mix　*n.* & *v.* 混合物，拌和物，新拌混凝土；（使）混合

thoroughness　*n.* 完全，充分

intermix　*v.* （使）混合

humidity　*n.* 湿度

substandard　*a.* 标准以下的，不合规格的

pronounced　*a.* 明显的，显著的

primitive　*a.* 原始的

antiquity　*n.* 古代

plastic　*a.* 可塑性的

deposit　*v.* 存放，堆积

member　*n.* 构件，部件

compressive strength　抗压强度

column　*n.* 圆柱，柱状物

arch　*n.* 拱形结构

brittle　*a.* 易碎的

tensile　*a.* 拉力的，能拉伸的

tensile strength　抗拉强度

footing　*n.* 基座，基础，基底

slab　*n.* 厚板，混凝土路面

massive structure　大体积结构

reinforced concrete　钢筋混凝土

reinforcement　*n.* 钢筋，增强，加固

retaining/retention wall　挡土墙

subject to　易受……的，有……倾向的

by weight　根据重量，依据重量

offset　*vt.* 弥补，抵消

reinforced concrete　钢筋混凝土

rod　*n.* 棒，杆

reinforcing steel　钢筋

deformation　*n.* 变形

interlocking　（使）相互扣住

integral　*a.* 完整的，整体的

formability　*n.* 可塑性，可成型的

ductility　*n.* 韧性，延展性

toughness   *n.* 坚韧,刚性
tank   *n.* 储水池
reservoir   *n.* 水库,蓄水池
a host of   一系列,许多
bending moment   弯矩
inevitable   *a.* 不可避免的
working load   工作荷载
elastic   *a.* 弹性的
elasticity   *n.* 弹性
amount to   合计,(总数)达到
strain   *n.* 应变
adverse   *a.* 不利的,相反的
unsightly   *a.* 难看的,不悦目的
corrosion   *n.* 侵蚀,腐蚀
induce   *v.* 引起,导致
prestress   *v. & n.* 预应力
strand   *n.* (绳,线等)股,缕,绞
equilibrium [复]equilibria 或 equilibriums   *n.* 平衡,平均,相称
flexure   *a.* 弯曲
service load   *n.* 工作荷载
thereby   *ad.* 因此,从而
prefabrication   *n.* 预制(构件)
mass-production   批量生产
sleeper   *n.* 枕木
in succession   接连地
hollow beams   空心梁
precast   *v.* 预制
lateral   *a.* 侧向的
prestressing cables   预应力筋
justify   *v.* 证明……是正当的

Text B   Proportioning, Batching and Mixing of Concrete

Proportioning of Concrete

The various components of a mix are proportioned so that the resulting concrete has adequate strength, proper workability for placing, and low cost. The last calls for use of the minimum amount of cement (the most costly of the components) which will achieve adequate properties. The better the gradation of aggregates, i.e., the smaller the volume of voids, the less cement paste is needed to fill these voids. In addition to the water required for hydration, water is needed for wetting the surface of the aggregate. As water is added, the plasticity and fluidity of the mix increases (i.e., its workability improves), but the strength decreases because of the larger volume of voids created by the free water. To reduce the free water while retaining the workability, cement must be added. Therefore, as for the cement paste, the water-cement ratio is the chief factor which controls the strength of the concrete. For a given water-cement ratio, one selects the minimum amount of cement, in sacks per cubic yard, which will secure the desired workability.

It has been customary to define the proportions of a concrete mix by the ratio, by volume or weight, of cement to sand to gravel, for example, 1 : 2 : 4. This method refers to the solid components only and, unless the water-cement ratio is specified separately, is insufficient to define the properties of the resulting concrete either in its fresh state or when set and hardened. For a complete definition of proportions it is now customary to specify, per 94 lb bag of cement, the weight of water, sand, and coarse aggregate. Thus, a mix may be containing (per 94 lb bag of cement) 45 lb of water, 230 lb of sand, and 380 lb of coarse aggregate. Alternatively, batch quantities are often defined in terms of the total weight of each component needed to make up 1 yd^3 of wet concrete, for example, 517 lb of cement, 300 lb of water, 1,270 lb of dry sand, and 1,940 lb of dry coarse aggregate.

Various methods of proportioning are used to obtain mixes of the desired properties from the cements and aggregates at hand. One is the so-called trial-batch method. Selecting a water-cement ratio from information, one produces several small trial batches with varying amounts of aggregate to obtain the required strength, consistency, and other properties with a minimum amount of paste. Concrete consistency is most frequently measured by the slump test . A metal mold in the shape of a truncated cone 12 in. high is filled with fresh concrete in a carefully specified manner. Immediately upon being filled, the mold is lifted off, and the slump of the concrete is measured as the difference in height between the mold and the pile of concrete. The slump is a good measure of the total water content in the mix and should be kept as low as is compatible with workability. Slumps for concretes in building construction generally range from 2 to 6 in.

The so-called ACI method of proportioning makes use of the slump test in connection with a set of tables which, for a variety of conditions (types of structures, dimensions of members, degree of exposure to weathering, etc.), permit one to estimate proportions which will result in the desired properties. These preliminary selected proportions are checked and adjusted by means of trial batches to result in concrete of the desired quality. Inevitably, strength properties of a concrete of given proportions scatter from batch to batch. It is therefore necessary to select proportions which will furnish an average strength sufficiently greater than the specified design strength for even the accidentally weaker batches to be of adequate quality.

In addition to the main components of concretes, admixtures are often used for special purposes. There are admixtures to improve workability, to accelerate or retard setting and hardening, to aid in curing, to improve durability, to add color, and to impart other properties. While the beneficial effects of some admixtures are well established, the claims for other should be viewed with caution. Air-entraining agents at present are the most important and most widely used admixtures. They cause the entrainment of air in the form of small dispersed bubbles in the concrete. These improve workability and durability, chiefly resistance to freezing and thawing, and reduce segregation during placing. They decrease density because of the increased void ratio and thereby decrease strength; however, this decrease can be partially offset by reduction of mixing water without loss of workability. The chief use of air-entrained concretes is in pavements, but they are also used for structures, particularly for exposed elements. Plasticizers and so-called superplasticizers are used increasingly, particularly for high strength concretes, because they permit significant water reduction while maintaining high slumps needed for proper placement and compaction of the concrete.

Batching and Mixing of Concrete

On all but the smallest jobs, batching is carried out in special batching plants. Separate hoppers contain cement and the various fractions of aggregate. Proportions are controlled, by weight, by means of manually operated or automatic dial scales connected to the hoppers. The mixing water is batched either by measuring tanks or by water meters. Admixtures are commonly provided in liquid form and may be dispensed into the mixer by weight or volume.

The principal purpose of mixing is to produce an intimate mixture of cement, water, fine and coarse aggregates, and possible admixtures of uniform consistency throughout each batch. This is achieved in machine mixers of the revolving-drum type. Minimum mixing time is 1 minute for mixers of not more than 1 yd^3 capacity, with an additional 15 second for each additional 1/2 yd^3. Mixing can be continued for a considerable time without adverse effect. This fact is particularly important in connection with ready-mixed concrete.

On large projects, particularly in the open country where ample space is available, movable mixing plants are installed and operated at the site. On the other hand, in construction under congested city conditions, on smaller jobs, and, frequently in highway construction, ready-mixed concrete is used. Such concrete is batched in a stationary plant and then hauled to the site in trucks in one of three ways:

(1) mixed completely at the stationary plant and hauled in a truck agitator;
(2) transit-mixed, i.e., batched at the plant but mixed in a truck mixer;
(3) mixed partially at the plant with mixing completed in a truck mixer.

Concrete should be discharged from the mixer or agitator within at most $1\frac{1}{2}$ hours after the water is added to the batch.

Words and Expressions

gradation   *n.* 分类,分级,级配

fluidity   *n.* 流动性
customary   *a.* 习惯的
cement paste   水泥浆
sack   *n. & v.* 袋,包,一袋,一包;灌袋
water-cement ratio   水灰比
batch   *n. & v.* 配料,定量混合物,一批;分批配料,分类,分组
trial-batch   (混凝土的)小量试拌
truncate   *v.* 截去……的顶端(末端)
cone   *n. & v.* 锥体,锥形;使成锥形,形成锥面
inevitably   *ad.* 不可避免地
scatter   *v.* 分散
furnish   *v.* 提供,供应
admixture   *n.* 掺合料,外加剂
plasticizer   *n.* 增塑剂,塑化剂,增韧剂
superplasticizer   *n.* 高效减水剂
hopper   *n.* 漏斗,底卸式车
fraction   *n.* 小部分
dial scale   带表盘秤
revolving-drum mixer   转筒式搅拌机
ready-mixed   预拌的
transit-mixed   运送时搅拌的
stationary   *a.* 固定的,定点的
haul   *n. & v.* 搬运,运输
discharge   *v.* 卸货,卸载
agitator   *n.* 搅拌器,搅拌机,混合器
batching plant   (混凝土)拌和厂

UNIT 12

Text A Conveying, Placing, Compacting, and Curing of Concrete

Conveying of Concrete

Conveying of most building concrete from the mixer or truck to the form is done in bottom-dump buckets or in wheelbarrows or buggies or by pumping through steel pipelines. The chief danger during conveying is that of segregation. The individual components of concrete tend to segregate because of their dissimilarity. In overly wet concrete standing in containers or forms, the heavier gravel components tend to settle, and the lighter materials, particularly water, tend to rise. Lateral movement, such as flow within the forms, tends to separate the coarse gravel from the finer components of the mix. The danger of segregation has caused the discarding of some previously common means of conveying, such as chutes and conveyor belts, in favor of methods which minimize this tendency.

Segregation

Segregation is a separation of the constituents of concrete so that their distribution ceases to be sufficiently uniform. It may be caused by a differential settlement of the aggregates in which the larger particles, or the heavier particles, travel faster down a pipe or slope, or settle faster in the concrete when they reach their final destination. Another type of segregation, which occurs particularly in wet mixes, results in the separation of the cement grout (that is, the cement and the water) from the aggregate and its formation in a layer on top of the concrete.

Segregation does not occur if the concrete mix is cohesive, but a workable mix is not necessarily cohesive. The slump test, the compacting factor test, and the Kelly Ball test all measure workability, but not cohesiveness. The flow test, however, does give an indication of cohesiveness. Air-entraining agents reduce the tendency towards segregation.

Evidently, great economies are achievable by the use of concrete pumps and slides, but these should be manipulated so that the concrete does not have to be moved sideways after it is deposited. The placement of concrete should be designed for it to travel the shortest possible distance to its final destination. Concrete should not be placed in layers thicker than 0.5 m to ensure that the layer below is still soft and that the two layers can be integrated by vibration. The vibration should be limited to the minimum required for consolidation, as excessive amounts produce segregation.

Bleeding

Closely related to segregation is bleeding, which is the collection of mixing water on the surface of freshly placed concrete; this water may carry some of the cement with it. It is the result of settlement of heavier particles and can be looked upon as a form of segregation. Bleeding can be expressed quantitatively as the total settlement per unit depth of concrete. There is an ASTM standard test for bleeding and the rate of bleeding.

Bleeding is not necessarily harmful if the water that collects on the surface evaporates before the concrete surface is given its final finish with a float. However, if the bleeding water brings cement to the surface, then a layer of set cement is formed on the surface; called laitance, this produces a dusty surface and a plane of weakness. If further concrete is to be placed on the top, then the laitance must be removed by brushing to ensure proper adhesion of the new concrete. Finishing with a wooden float, instead of a steel float, avoids overworking of the surface and bringing an excess of cement to the top.

Bleeding water may also become trapped under large aggregate particles or under reinforcing bars, where it forms capillary voids on evaporation; these may have an adverse effect on durability, particularly on resistance to frost.

The tendency to bleeding can be reduced by the use of air-entraining agents, by the use of finer cement and a greater proportion of very fine aggregate, by a decrease in the water content, and by a decrease in the water/cement ratio (by reducing water content or increasing cement content).

Placing and Compacting of Concrete

Placing is the process of transferring the fresh concrete from the conveying device to its final place in the forms. Prior to placing, loose rust must be removed from reinforcement, forms must be cleaned, and hardened surfaces of previous concrete lifts must be cleaned and treated appropriately. Placing and compacting are critical in their effect on the final quality of the concrete. Proper placement must avoid segregation, displacement of forms or of reinforcement in the forms, and poor bond between successive layers of concrete. Immediately upon placing, the concrete should be compacted by means of hand tools or vibrators. Such compacting prevents honeycombing, assures close contact with forms and reinforcement, and serves as a partial remedy to possible prior segregation. Compacting is achieved by hand tamping with a variety of special tools, but now more commonly and successfully with high frequency, power-driven vibrators. These are of the internal type, immersed in the concrete, or of the external type, attached to the forms. The former are preferable but must be supplemented by the latter where narrow forms or other obstacles makes immersion impossible.

Curing of Concrete

Fresh concrete gains strength most rapidly during the first few days and weeks. Structural design is generally based on the 28-day strength, about 70 percent of which is reached at the end of the first week after placing. The final concrete strength depends greatly on the conditions of moisture and temperature during this initial period. The maintenance of proper conditions during this time is known as curing. Thirty percent of the strength or more can be lost by premature drying out of the concrete. To prevent such damage, concrete should be protected from loss of moisture for at least 7 days and, in more sensitive work, up to 14 days. When high early strength cements are used, curing periods can be cut in half.

Both high and low temperatures adversely affect the hydration of the cement. Special precautions are necessary when placing concrete in hot weather. The temperature of new concrete should be kept below 32 ℃ and preferably below 30 ℃. This may be accomplished by mixing crushed ice with the mixing water and by keeping all accessory materials in the shade. The heat of hydration produced by the cement can be reduced by using a low-heat cement and by using special care in curing to prevent evaporation of the mixing water.

Special precautions are also necessary in cold weather, that is, when the temperature falls below 4 ℃. The concrete should be kept above a temperature of 15 ℃. This may be accomplished by heating the mixing water and the aggregates; however, the water should not be heated above 65 ℃ to avoid a flash set of the concrete. Cement with a higher content of calcium aluminate and of tricalcium silicate should be used to generate a higher rate of heat development. Calcium chloride as an accelerator increases the rate of hydration and thus generates further heat; in addition, it turns the mixing water into a salt solution and thus lowers its freezing point below that of pure water. The concrete must be further protected against frost during curing.

Curing is the term given to the protection of the concrete during the early stages of hardening, when it requires additional moisture for the continuing hydration of the cement, and to its protection from cold during the night and from heat during the day. The concrete surface can be covered with sand or burlap, which is kept moist by watering or spraying. Precast concrete units and concrete slabs may be covered with water.

Curing can be achieved by keeping exposed surfaces continually wet through sprinkling, and ponding. Alternatively, the concrete can be covered with a curing membrane, which provides a physical barrier to the evaporation of its water. This is applied either as a sheet or as a liquid that dries within a few hours to a continuous, adhesive film; a dye is sometimes added to facilitate uniform coverage. Polyvinyl chloride, polyethylene (about 0.2 mm thick), or waterproof building paper provide suitable sheet materials. Liquid curing compounds include various wax and oil emulsions as well as various plastics.

Precast concrete units are sometimes steam-cured or autoclaved (cured with high-pressure steam), a treatment that increases their strength and makes it possible to handle and transport them much sooner, thus greatly reducing the amount of storage space required.

In addition to improved strength, proper moist curing provides better shrinkage control. To protect the concrete against low temperatures during cold weather, the mixing water and, occasionally, the aggregates are heated, temperature insulation is used where possible, and special admixtures, particularly calcium chloride, are employed. When the air temperature is very low, external heat may have to be supplied in addition to insulation.

Words and Expressions

convey　*v.* 输送，运送
conveyor belt　传送带
bucket　*n.* 吊斗，挖斗
wheelbarrow　*n.* 手推车，独轮小车
buggy　*n.* 手推车，小斗车
segregation　*n.* 离析
dissimilarity　*n.* 不相似，不同
overly　*ad.* 过度地
discard　*v.* 放弃，抛弃
chute　*n.* 斜槽，溜槽；斜管
differential　*a.* 区别的，差别的
bleeding　*n.* （混凝土表面）泛出水泥浮浆，泌水，分泌
grout　*n. & v.* 灰浆，水泥浆；灌浆，浆砌
cohesive　*a.* 黏聚的，黏结的，有附着性的
slide　*n.* 滑槽
manipulate　*v.* 处理
deposit　*v. & n.* 浇筑；沉积物
consolidation　*n.* 压实，捣实
subsidence　*n.* 沉降，沉陷
place　*v.* 浇筑（混凝土），放置
lift　*n.* （混凝土）浇筑层
displacement　*n.* 变位，移动，偏移
vibrator　*n.* 震动器，振捣器
compact　*v.* 压实，夯实，捣实，振捣
honeycomb　*n. & a.* 蜂窝结构；蜂窝状的
tamping　*n.* 夯实，捣固，捣实

float   n. 镘刀,用镘刀抹平
laitance   n. 水泥浮浆,混凝土面上的沫状物
trap   v. 使受限制,止住
capillary   n. 毛细管,毛细现象
capillary void   毛细孔
adverse   a. 不利的,有害的;相反的
immerse   v. 浸入
supplement   v. 补充
obstacle   n. 障碍
premature   a. 过早的
sprinkling   n. 喷洒,撒布,洒水
ponding   n. 积水,蓄水
retard   v. 阻止
hydration   n. 水化,水化作用
flash set   急凝,瞬时凝结
generate   v. 释放
burlap   n. 粗麻布,麻袋
precast   v. 预制,预浇
membrane   n. 膜
calcium chloride   氯化钙
polyvinyl chloride   n. 聚氯乙烯
polyethylene   n. 聚乙烯
emulsion   n. 乳胶体,乳剂,乳液

Text B   Concrete Properties

Workability of Concrete

Fresh concrete should be such that it can be transported, placed, compacted, and finished without harmful segregation. A proper mix should maintain its uniformity inside the forms and should not bleed excessively. All these aspects of fresh concrete—those associated with several tasks from selection of materials to finishing—are collectively called the workability. Thus workability can be defined as the ease with which a fresh concrete mix can be handled from the mixer to the final structure. At present there does not exist a procedure to measure the workability quantitatively. But a nonworkable mixture can easily be identified from the inability to satisfy one or more of the concrete tasks: mixing, transporting, compacting, and finishing.

The three primary characteristics of workability are: consistency, mobility, and compactability.

In addition, for some structures, such as floor slabs, finishability is also a measure of workability.

Consistency is a measure of concrete wetness or fluidity, which depends on the mix proportions and properties of the ingredients. It is generally measured with a slump test. The test is also used to measure the characteristics of workability, although it only measures (indirectly) the ease with which concrete can be placed, compacted, and finished without harmful segregation. Thus the slump or slump test can be used to measure changes in workability for relative comparison of workability between different mixtures.

Slump Test

The slump test was developed by Chapman in the United State in 1913. It consists of a metal slump cone having a bottom diameter of 200 mm and a top diameter of 100 mm. The height of the cone is 300 mm. The cone (whose inside surface is dampened) is placed on a smooth, flat, nonabsorbent surface and is filled with fresh concrete while it is held firmly in place by standing on the foot pieces. It is filled to about one-third of its height, which is then tamped 25 times with a 16-mm (600-mm long) tamping rod (Fig. 1). The process is repeated two more times when the cone is filled, and the top is struck off with the tamping rod (placed horizontally) so that the cone is filled exactly. The mold is then removed by raising it vertically immediately after filling. The concrete will subside, and the number of centimeters the center of the cone settles or slumps is measured. This is called the slump. A very dry mix will have zero slump, whereas a very wet mix collapses completely, producing unreliable value of slump.

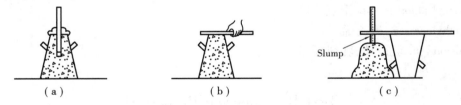

Fig. 1 Slump test

(a) Fill the cone using standard procedure; (b) Strike off excess;

(c) Raise the mold and measure the displacement of the original center of the specimen

The slump depends on the ingredients, amounts of mixing water, and addition of admixtures. The value of slump also changes with temperature and time after mixing (owing to the hydration process and the evaporation of water).

Compressive Strength of Concrete

The compressive strength is the most important property of hardened concrete and is generally considered in the design of concrete mixtures. Although concrete can be manufactured to have a compressive strength as high as 82.7 MPa, in ordinary construction a strength in the range 20.7 to 41.4 MPa can be expected. Cast-in-place construction generally uses concrete of strength closer to the low range, whereas precast construction, which routinely follows a rigorous quality control program, makes use of concrete of strength closer to the high range.

It is customary to estimate the properties of concrete in the structure from compression tests on specimens made from fresh concrete as it is placed and cured in the standard condition. Since the compressive strength is affected by many variables which include the environmental factors and the curing conditions, the actual strength of the concrete (in structure) will not be the same as the strength of the specimen.

A number of factors, such as amount of cement, amount of water, types of ingredients, mix proportions, curing, temperature, age, size and shape of specimen, and test conditions affect the compressive strength.

Size and Shape of Aggregates

Strength of concrete of a given consistency increases with the increase in fineness modulus of fine aggregate. Fineness modulus indicates the average size of particles in the graded aggregate. A high number means coarser gradation, and a lower number reflects finer gradation. Fineness modulus of graded sand for use in concrete generally lies between 2.3 and 3.1. Natural sand of fineness modulus less than 1.5 is seldom found. It is found that as the gradation becomes coarser, compressive strength increases.

Strength of concrete may also be affected by the type and size of coarse aggregates. Because of their surface texture and particle shape, limestone and granite aggregates may exhibit compressive strength up to 20% higher than river gravel using the same water/cement ratio. The strength of lightweight concrete is lower than that of normal-weight concrete of equal consistency.

The larger the maximum size of coarse aggregate, the lower the surface area and therefore the lower the water requirement for a set consistency. With a lower water/cement ratio, the strength of concrete is increased. However, the use of larger aggregate per se decreases the compressive strength. This is probably due to less surface, which lowers the bond.

Rich mixes (high cement content) are adversely affected by the use of large aggregate. In lean mixes, however, increase in aggregate size lowers the water requirement and may actually increase the compressive strength. Typically, the smallest coarse aggregate produces the highest strength for a given water/cement ratio. With larger aggregate, high strength can be obtained with the use of superplastcizing admixture.

Water/Cement Ratio

For a given cement content, the maximum strength occurs in a mix that contains only sufficient water for complete hydration. However, this mix may be dry and harsh, and it may be difficult to obtain homogeneity. When the cement content remains constant, the strength of concrete decreases with the increase of the amount of mixing water.

Although the water/cement ratio is an important property in the design of concrete mixtures, it should be noted that the same strength cannot be expected with different aggregate, even when the water/cement ratio is kept the same. As an example, for the same water/cement ratio, mortar, which does not have coarse aggregate, is stronger than concrete.

Voids

The increase of water content increases the voids in concrete, which decreases the durability, watertightness, and the compressive strength. The water content should be enough to guarantee complete hydration of Portland cement. Any excess water added to the mix increases the void content, which in turn decreases the quality of concrete. But it should be recognized that insufficient amount of cement means an inadequate amount of binder, and the resulting concrete is definitely of lower quality.

In summary, although the water/cement ratio is a convenient way to do mix design, good-quality concrete requires a sufficient amount of cement, well-graded aggregates, and a minimum amount of mixing water.

Age and Curing

Strength of concrete is affected by age and curing. The effects of length of curing (age) and the temperature on the compressive strength are as shown in Fig. 2 and Fig. 3. Plastic concrete (or fresh concrete) has negligible strength. Strength of concrete at 1 day is about 10% to 20% of its 28-day strength. The 7-day strength is about 70% to 80% of the 28-day strength. It can be seen from Fig. 2 that moist-cured specimens continue to gain strength even after several years, but on the other hand, specimens that are left to dry cease to hydrate and gain strength. Improvement in strength beyond a year is very little. The increase of water temperature, both at mixing and curing, makes the strength gain faster. Using steam to cure concrete has a dramatic effect on the strength gaining.

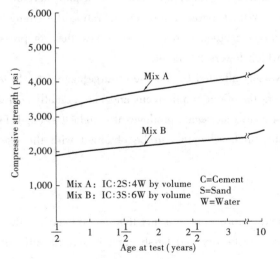

Fig. 2   Effect of age on compressive strength of concrete   Fig. 3   Effect of curing temperature on f'_c

Air Entrainment

Concrete made with air-entraining chemicals, or air-entrained concrete, usually has a lower compressive strength than ordinary concrete. Entrained air in the amount of 3% to 8%, without adjustment of the water/cement ratio, will reduce the compressive strength by about 5% for each 1% of added air. But in practice, the water/cement ratio and sand content of air-entrained concrete will be decreased to obtain the same strength as that of normal concrete.

Compression Test

In the United States, the compressive strength of concrete is determined from compression test on cylindrical specimens 6 in. in diameter and 12 in. in height. Empty metal cylinders are filled with fresh concrete using a standard procedure. After 24 hours the specimens are taken out of the molds and moist cured for 28 days. At the end of the curing period they are capped and tested in a moist condition. The failure load divided by the cross-sectional area is called 28-day cylinder compressive strength (f'_c).

$$f'_c = \frac{P}{(\pi/4)d^2}$$

where P the failure load and d the diameter of the cylinder.

The compressive strength thus determined is found to depend on the size of the specimen, the shape of the specimen, and the moisture condition. The greater the ratio of height to diameter, the lower is the measured compressive strength. For specimens of height equal to twice the diameter, the compressive strength decreases with the increase of diameter. A 4-in.-diameter cylinder (8 in. in height) exhibits approximately 5% higher strength than that of a 6-in.-diameter cylinder (12 in. in height).

The moisture content of specimens affects the compressive strength. Air-dried specimens (at the time of testing) are shown to possess more compressive strength than that of saturated specimens, on the order of 20% to 25%. The strength is also affected by the speed of testing—a slower rate will show a lower strength. In the laboratory the rate of loading is adjusted so that failure takes place within 2 to 3 minutes.

Words and Expressions

concrete finish   混凝土收面,混凝土饰面
segregation   *n.* 离析
slump test   坍落度试验
a metal slump cone   金属坍落度筒
dampen   *v.* 湿润,使潮湿
tamp   *v.* 夯实
horizontally   *ad.* 水平地
vertically   *ad.* 垂直地
subside   *v.* 下沉,沉降,下陷
cast-in-place   现浇
routinely   *ad.* 常规地

rigorous　*a.* 严格的
fineness modulus　细度模数
gradation　*n.* 级配
graded aggregate　级配集料
granite　*n.* 花岗石
river gravel　河卵石
set　*a.* 固定的
per se　*ad.* 本身,本质上
lean　*a.* 贫瘠的
superplastcizing admixture　高效塑化剂
homogeneity　*n.* 均匀性
harsh　*a.* 粗糙的
durability　*n.* 耐久性
watertightness　*n.* 水密性,不透水性
cylindrical　*a.* 圆柱形的
cylinder　*n.* 圆柱
moisture content　含水量

UNIT 13

Text A Durability of Concrete

Besides its ability to sustain loads, concrete is also required to be durable. The durability of concrete can be defined as its resistance to deterioration resulting from external and internal causes. The external causes include the effects of environmental and service conditions to which concrete is subjected, such as weathering, chemical actions and wear. The internal causes are the effects of salts, particularly chlorides and sulphates, in the constituent materials, interaction between the constituent materials, such as alkali-aggregate reaction, volume changes, absorption and permeability.

In order to produce a durable concrete care should be taken to select suitable constituent materials. It is also important that the mix contains adequate quantities of materials in proportions suitable for producing a homogeneous and fully compacted concrete mass.

Weathering

Deterioration of concrete by weathering is usually brought about by the disruptive action of alternate freezing and thawing of free water within the concrete and expansion and contraction of the concrete, under restraint, resulting from variations in temperature and alternate wetting and drying.

Damage to concrete from freezing and thawing arises from the expansion of pore water during freezing; in a condition of restraint, if repeated a sufficient number of times, this results in the development of hydraulic pressure capable of disrupting concrete. Road kerbs and slabs, dams and reservoirs are very susceptible to frost action.

The resistance of concrete to freezing and thawing can be improved by increasing its impermeability. This can be achieved by using a mix with the lowest possible water-cement ratio compatible with sufficient workability for placing and compacting into a homogeneous mass. Durability can be further improved by using air-entrainment, an air content of 3 to 6 percent of the volume of concrete normally being adequate for most applications. The use of air-entrained concrete is particularly useful for roads where salts are used for deicing.

Chemical Attack

In general, concrete has a low resistance to chemical attack. There are several chemical agents which react with concrete but the most common forms of attack are those associated with leaching, carbonation, chlorides and sulphates. Chemical agents essentially react with certain compounds of the hardened cement paste and the resistance of concrete to chemical attack therefore can be affected by the type of cement used. The resistance to chemical attack improves with increased impermeability.

Wear

The main causes of wear of concrete are the cavitation effects of fast-moving water, abrasive material in water, wind blasting and attrition and impact of traffic. Certain conditions of hydraulic flow result in the formation of cavities between the flowing water and the concrete surface. These cavities are usually filled with water vapour charged with extraordinarily high energy and repeated contact with the concrete surface results in the formation of pits and holes, known as cavitation erosion. Since even a good-quality concrete will not be able to resist this kind of deterioration the best remedy is therefore the elimination of cavitation by producing smooth hydraulic flow. Where necessary, the critical areas may be lined with materials having greater resistance to cavitation erosion.

In general, the resistance of concrete to erosion and abrasion increases with increase in strength. The use of a hard and tough aggregate tends to improve concrete resistance to wear.

Alkali-Aggregate Reactions

Certain natural aggregates react chemically with the alkalis present in Portland cement. When this happens these aggregates expand or swell resulting in cracking and disintegration of concrete.

Volume Changes

Principal factors responsible for volume changes are the chemical combination of water and cement and the subsequent drying of concrete, variations in temperature and alternate wetting and drying. When a change in volume is resisted by internal or external forces this can produce cracking, the greater the imposed restraint, the more severe the cracking. The presence of cracks in concrete reduces its resistance to the action of leaching, corrosion of reinforcement, attack by sulphates and other chemicals, alkali-aggregate reaction and freezing and thawing, all of which may lead to disruption of concrete. Severe cracking can lead to complete disintegration of the concrete surface particularly when this is accompanied by alternate expansion and contraction.

Volume changes can be minimized by using suitable constituent materials and mix proportions having due regard to the size of structure. Adequate moist curing is also essential to minimize the effects of any volume changes.

Permeability and Absorption

Permeability refers to the ease with which water can pass through the concrete. This should not be confused with the absorption property of concrete and the two are not necessarily related. Absorption may be defined as the ability of concrete to draw water into its voids. Low permeability is an important requirement for hydraulic structures and in some cases water-tightness of concrete may be considered to be more significant than strength although, other conditions being equal, concrete of low permeability will also be strong and durable. A concrete which readily absorbs water is susceptible to deterioration.

Concrete is inherently a porous material. This arises from the use of water in excess of that required for the purpose of hydration in order to make the mix sufficiently workable and the difficulty of completely removing all the air from the concrete during compaction. If the voids are interconnected, concrete becomes pervious although with normal care concrete is sufficiently impermeable for most purposes. Concrete of low permeability can be obtained by suitable selection of its constituent materials and their proportions followed by careful placing, compaction and curing. In general for a fully compacted concrete, the permeability decreases with decreasing water-cement ratio. Permeability is affected by both the fineness and the chemical composition of cement. Coarse cements tend to produce pastes with relatively high porosity. Aggregate of low porosity are preferable when concrete with a low permeability is required. Segregation of the constituent materials during placing can adversely affect the impermeability of concrete.

Words and Expressions

sustain   *v.* 承受
durability   *n.* 耐久性,耐用年限
deterioration   *n.* 变坏,变质,损坏,损伤
chloride   *n.* 氯化物,漂白粉
sulphate   *n.* 硫酸盐
alkali-aggregate reaction   碱集料反应
absorption   *n.* 吸收;吸水性
permeability   *n.* 渗透性,透气性
homogeneous   *a.* 均质的,均匀的
weathering   *n.* 风化,自然老化,大气侵蚀
disruptive   *a.* 摧毁的,破坏的
thaw   *v.* 融化,解冻
restraint   *n.* 限制,抑制
compatible with   与……共存的
entrain   *v.* 携带,使(空气)以气泡状存于混凝土中
air-entrainment   引气剂

air-entrained concrete　加气混凝土
deice　*v.* 化雪
leaching　*n.* 浸出,浸析作用,溶析
carbonation　*n.* 碳化作用
cavity　*n.* 空腔,空穴,孔穴,洞穴
cavitation　*n.* 气蚀,空蚀,气穴
abrasive　*a.* 研磨的
blasting　*n.* 爆破
attrition　*n.* 磨损,磨耗,损耗
hydraulic　*a. & n.* 水力的,水压的;水力
remedy　*n.* 补救
impose　*v.* 施加
disruption　*n.* 分裂,瓦解
due　*a.* 适当的
have regard to　考虑
impermeability　*n.* 不渗透性,防水性,气密性
hydraulic structure　水工建筑物
inherently　*ad.* 固有地
porous　*a.* 多孔的,可透水的
pervious　*a.* 透水的,透光的,有孔的,能通过的

Text B　Durability of Building Materials

Some building materials are likely to last as their building does while others will require periodic replacement. In life-cycle costing, the cost of such replacement is taken into account, but it is a matter of opinion whether the choice of a material should be based on its initial cost or its life-cycle cost.

Some materials—for example, glazed tile and stainless steel sheet—are extremely durable; once installed in a new building, they may not require renewal during its entire life. Other materials, such as paint and galvanized steel sheet, will need renewal several times during its life. Is it then cheaper to use galvanized steel or stainless steel for the rain gutters? Is it cheaper to use glazed tiles or painted plaster?

In estimating a building on present cost, the maintenance is ignored, or at least not formally considered. In life-cycle costing, the capitalized cost of the maintenance during the entire life of the building is estimated to obtain the total cost. Many materials that require frequent renewal or maintenance have a far higher life-cycle cost than others that cost much more initially, but require no maintenance. Thus, it is likely that stainless-steel gutters and glazed tiles will have a lower life-cycle cost but galvanized-steel gutters and painted plaster the lower initial cost.

High or low temperatures and high or low humidity in themselves do not cause significant damage to building materials, but changes in temperature and humidity do. Thermal expansion and contraction can cause cracks in brittle materials and so can moisture movement. The alternate wetting and drying that takes place during and after rain can also cause appreciable damage, particularly to materials that absorb some of the water.

Some bricks and natural stones contain soluble salts that rain water can bring to the surface, where they form efflorescence. This can usually be removed with a brush. Efflorescence may also occur on concrete containing unsuitable aggregates.

Most climatic zones experience a substantial number of days when the temperature at night drops below the freezing point (0 ℃ or 32 ℉) and rises above it during daytime. Since ice has a greater volume than water, these cycles of freezing and thawing can be very damaging to materials that are both porous and brittle.

The spectrum of solar radiation includes both infrared (heat) radiation with a longer wavelength and ultraviolet radiation with a shorter wavelength; both overlap with visible light. Heat radiation can produce overheating of some materials. Thus, bituminous flat roofs are usually covered either with a thin layer of light-colored paint or white stone chips to reflect some of the radiation or with a thicker layer of gravel. Ultraviolet radiation is a major cause of deterioration of certain plastics, and some of these cannot be used if they are to be exposed to sunlight.

Sunlight fades certain pigments, mostly those of organic origin, and these should not be used externally either in paints or for coloring anodized aluminum. Sunlight is also a major cause of the breakdown of paint film, which is further accelerated by the thermal movement that occurs.

Iron and steel corrode in the presence of moisture by forming rust, the common term for hydrated iron oxide ($2Fe_2O_3 \cdot 3H_2O$), unless the climate is extremely dry.

The reinforcement in concrete must be protected by an adequate cover of concrete; if this is insufficient, spalling may occur.

Erosion of external walls is caused by wind-driven particles of sand. It occurs comparatively rarely since it requires high wind velocities or the formation of eddies, a supply of sand or dust, and soft material in the wall. The main damage is to old buildings that contain deteriorating stone or brick. Abrasion is the damage caused by fine solid particles to floor surfaces. Wear is a more complex phenomenon that is caused partly by abrasion, but also by compression and by impact. Carpeting, for example, is frequently damaged more by the permanent impression made by heavy pieces of furniture than by abrasion. It is relatively simple to test abrasion resistance, however, and many machines for testing accelerated wear are simply abrasion-testing machines.

Concrete is one of the best floor surfaces for industrial use, and it can be further improved by using a hard aggregate at least for the surface layer. Cast iron tiles may be appropriate for conditions of extreme wear.

Quarry tiles, which are hard-burned, unglazed clay tiles, provide one of the hardest wearing surfaces for commercial and domestic buildings. Tile and stone floors provide heat storage for passive solar design.

Carpeting is used increasingly, both for offices and for homes, because its relative cost has been greatly reduced by modern manufacturing processes that enable the entire floor to be covered wall to wall. Carpet provides a comfortable walking surface; it is easier to clean than a hard surface; and it is a good thermal insulator and an excellent absorber of impact sound.

Words and Expressions

glaze　　*n.* 釉面,上釉,半透明涂层
stainless steel　　不锈钢
renewal　　*n.* 翻新
galvanize　　*v.* 电镀,镀锌于
plaster　　*n.* 灰泥,灰浆
thermal　　*a.* 热的,热量的
efflorescence　　*n.* 风化
spectrum　　*n.* 谱,频谱,领域,范围,各种各样的
infrared　　*a. & n.* 红外的,产生红外辐射的;红外线
ultraviolet　　*a. & n.* 紫外的,紫外线的;紫外线辐射
radiation　　*n.* 发射,辐射,照射,放射线
overlap　　*v.* 重叠,交搭
fade　　*v.* 衰减,(使)褪色,逐渐消失
pigment　　*n.* 颜料
anodized　　*a. & n.* 受过阳极化处理的(金属表面),阳极氧化膜
breakdown　　*n.* 剥落
corrode　　*v.* 腐蚀,侵蚀
hydrate　　*n. & v.* 水合物;(使)成水合物
spall　　*v. & n.* 剥落,散裂,裂开;裂片,碎片
erosion　　*n.* 侵蚀,腐蚀,冲刷,冲蚀
abrasion　　*n.* 磨损,磨耗,磨蚀
quarry　　*n.* 方形砖,方形瓦
eddy　　*n.* (水,风,尘等的)涡流,旋涡运动
carpet　　*n. & v.* 地毯,磨耗层;铺毡,铺毯,铺层
quarry tile　　黏土瓦
domestic　　*a.* 家庭的,民用的
insulator　　*n.* 绝缘(热)体

UNIT 14

Text A Rigid Pavement

Rigid pavement is made up of Portland cement concrete and may or may not have a base course between the pavement and the subgrade. It is different from flexible pavement, which includes several layers of structural components of the pavement. In the rigid pavement, the concrete, exclusive of the base, is referred to as the pavement. The term "rigid" implying that pavements constructed of this material possess a certain degree of "beam strength" that permits them to span or "bridge over" some minor irregularities in the subgrade or subbase on which they rest. Thus minor defects or irregularities in the supporting foundation layer may not be reflected in the surface course, although of course, defects of this type are certainly not desirable, as they may lead to failure of the pavement through cracking, breaking, or simpler distress.

The essential difference between the rigid pavement and flexible pavement is the manner in which they distribute the load over the subgrade. The rigid pavement, because of its rigidity and high modulus of elasticity, tends to distribute the load over a relatively wide area of soil; thus, a major portion of the structural capacity is supplied by the slab itself. The major factor considered in the design of rigid pavements is the structural strength of the concrete. For this reason, minor variations in subgrade strength have little influence upon the structural capacity of the pavement.

Base course are used under rigid pavements for various reasons, including (1) control pumping, (2) control of frost action, (3) drainage, (4) control of shrink and swell of subgrade, and (5) expedition of construction. The base course will lend some structural capacity to the pavement; however, its contribution to the load-carrying capacity is relatively minor.

When properly designed and constructed, concrete roads and streets are capable of carrying almost unlimited amounts of any type of traffic with ease, comfort, and safety. Surfaces of this type are smooth, dust-free, and skid-resistant, having a high degree of visibility for both day and night driving and generally having low maintenance costs. They are economical in many locations because of their low cost of maintenance and their relative permanence. They are, of course, classed as high-type pavements. The principal use of surfaces of this type has been in the construction of heavily traveled roads and city streets, including those in residential, business, and industrial areas. It is the standard material for urban expressways, even in states where asphalt surfaces are widely used. A wearing surface of Portland cement concrete usually consists of a single layer of uniform cross section that has a thickness of 6 to 11 in. and that may not require a separate base course, often being constructed directly on a prepared subgrade or subbase. A new concrete base may be constructed to serve as a support for one of the several types of bituminous wearing surfaces. Old concrete pavements have been extensively used as bases for new bituminous wearing surfaces in many areas.

Words and Expressions

rigid pavement　刚性路面
Portland cement concrete　波特兰水泥混凝土
component　*n.* 组成成分
exclusive of　除,不计算……在内
be referred to as　被称为
defect　*n.* 缺陷
modulus of elasticity　弹性模量
slab　*n.* （水泥混凝土）板
shrink and swell　收缩和膨胀
ease　*n.* 舒适

Text B　Flexible Pavement

A highway pavement is a structure consisting of superimposed layers of selected and processed materials whose primary function is to distribute the applied vehicle loads to the subgrade[①]. The ultimate aim is to ensure that the transmitted stresses are sufficiently reduced that they will not exceed the supporting capacity of the subgrade. Two types of pavement are generally recognized as serving this purpose—flexible pavements and rigid pavements.

A flexible pavement is a pavement structure that maintains intimate contact with, and distributes loads to, the subgrade; it depends upon aggregate interlock, particle friction, and cohesion for its stability. The strength of the subgrade is a major factor controlling the design of a flexible pavement. When the subgrade deflects, the overlying flexible pavement is assumed to deform to a similar shape and extent. For this the assumed basic design criterion is that a depth of pavement is required that will distribute the applied surface load through the various pavement layers to the subgrade so that the subgrade is not over-stressed.

UNIT 14

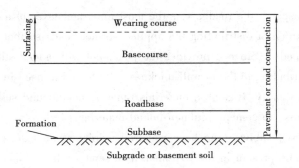

Fig. 1 The basic structural cross-section of a flexible road

Whether it is a pavement for an expensive motorway or a simple country road, the basic structural cross-section of a flexible road is essentially illustrated in Fig. 1; it is composed of several distinct layers that make up the pavement superimposed on the subgrade in the manner indicated. The intersection of the subgrade and the pavement is known as the formation.

The subgrade is normally considered to be the in situ[2] soil over which the highway is being constructed. It should be quite clear, however, that the term subgrade is also applied to all native soil materials exposed by excavation and to excavated soil that may be artificially deposited to form a compacted embankment. In the latter case, the added material is not considered to be part of the road structure itself but part of the foundation of the road.

The uppermost layer of a flexible pavement is called the surface course. The highway materials used in a surface course can vary from loose mixtures of soil and gravel to the very high-quality bituminous mixtures. The choice of materials used in any particular situation depends in most countries upon the quality of service required of the highways[3].

If a surface course is composed of bituminous materials, as is the normal practice for flexible pavements in Britain, it may consist of a single homogeneous layer or, in the higher-quality roads, of two distinct sub-layers known as a wearing course and a basecourse. The wearing course provides the actual surfacing on which the vehicles run, whilst the basecourse acts as a regulating layer to provide the wearing course with a better riding quality. The basecourse is normally composed of a more pervious material than the wearing course.

The primary function of the surface course, and especially of its wearing course component, is to provide a safe and comfortable riding surface for traffic. It must also withstand the most concentrated stresses due to traffic, and protect the pavement layers beneath from the effects of the natural elements.

Bituminous surfaces are generally expected: (a) to contribute to the structural strength of the pavement, (b) to provide a high resistance to plastic deformation and resistance to cracking under traffic, and (c) to maintain such desirable surface characteristics as good skid-resistance, good drainage, and low tyre noise.

Roadbase must not be confused with the basecourse, which is an integral part of the surface course. One is a sub-layer within the bituminous surfacing, while the other is normally the thickest element of the flexible pavement on which the surfacing rests.

From a structural aspect, the roadbase is the most important layer of a flexible pavement. It is expected to bear the burden of distributing the applied surface loads so that the bearing capacity of the subgrade is not exceeded. Since it provides the pavement with added stiffness and resistance to fatigue, as well as contributing to the overall thickness, the material used in a roadbase must always be of a reasonably high quality. Roadbase materials range from unbound soils and/or aggregates to chemically stabilized soils, to cement/bitumen-bound materials.

In its simplest sense, subbase can be considered merely as an extension of the roadbase; in fact, it may or may not be present in the pavement as a separate layer. Whether or not it is utilized in a pavement depends upon the purpose for which it is to be used. Its function can be examined from a number of aspects, as follows.

(1) As a structural member of the pavement the subbase helps to distribute the applied loads to the subgrade. The subbase material must always be significantly stronger than the subgrade material and capable of resisting within itself the stresses transmitted to it via the roadbase.

(2) A coarse-grained material may be used in the subbase to act as a drainage layer, i.e. to pass to the highway drainage system any moisture which falls during construction or which enters the pavement after construction. The quality of the material used must be such that the free-drainage criterion of the subbase is always met; in certain instances, this may require a dual-layer subbase, i.e. an open-graded layer with a protective filter.

(3) On fine-grained subgrade soils a granular subbase may be provided: (a) to carry constructional traffic and act as a working platform on which subsequent layers can be constructed, (b) to act as a cutoff blanket and prevent moisture from migrating upward from the subgrade, or (c) to act as a cutoff blanket to prevent the infiltration of subgrade material into the pavement.

Words and Expressions

flexible   *a.* 柔性的;灵活的;可塑造的
flexible pavement   柔性路面
superimpose   *v.* 叠加,放在上面
process   *v.* 加工
transmit   *v.* 传送,传导,传达
rigid   *a.* 刚性的,坚硬的
interlock   *n.* 连接;嵌锁,锁结
aggregate interlock   集料嵌锁
particle   *n.* 颗粒,微粒;粒子
particule friction   微粒摩擦
cohesion   *n.* 黏力,凝聚力,内聚
deflect   *v.* (使)偏斜,(使)偏差;挠曲;变位
deform   *v.* (使)变形
overstress   *v.* 应力过度

formation   n. 路基面；构造；构成
subbase   n. 底基层，副基层
deposite   v. 使沉积，使淤积
uppermost layer   最高层，表层
bituminous   a. & n. 沥青的；沥青
homogeneous   a. 同种的，同质的；均匀的
basecourse   n. 下面层；基层
whilst   conj. (=the while) 与此同时
regulating layer/course   整平层
pervious   a. (可)透水的；有孔的
cracking   n. 开裂，破裂；裂缝，裂纹
skid   n. 滑；(汽车)打滑
skid-resistance   抗滑(性)
roadbase   n. (道路)基层
confuse   v. 使混乱，混淆
integral   a. 构成整体所必要的，组成的
stiffness   n. 刚度
fatigue   n. 疲劳，疲乏
unbound   a. 未结合的，自由的
utilize   v. 利用
coarse-grained   粗颗粒的
dual-layer   双层
open-graded   开级配的
filter   n. 过滤器，滤波器；滤机
granular   a. 黏状的，黏料的
cutoff   n. 隔离，切断
cutoff blanket   隔水层，截水层
infiltration   n. 渗入，渗透
stabilize   v. 使稳定，坚固，不动摇等
via   prep. 经过，取道
surface course   表层
bitumen-bound   沥青结合的
free-draining   自由排水
excavation   n. 挖土，挖掘
embankment   n. 路堤
clue   n.& v. 线索，思路；提示

Notes

①这是关系代词 who 的所有格,也可称为关系形容词。文中的关系从句是 whose primary function is to distribute the applied vehicle loads to the subgrade,其主要作用是将车辆的外加荷载分散(送)到路基去。关系形容词 whose 既可指代人,也可指代物。例如:Every profession or trade, every art, and every science has its technical vocabulary, whose function is partly to designate things or processes which have no names in ordinary English.

②in situ 意为"现场,就地",此处作定语。

③required of the highway,如同本句的 used in any particular situation 一样,是过去分词短语作后置定语,可将其扩展成定语从句 that is required of the highway。Require/ask/demand of sb./ sth.意为"要求某人/某物什么"。例如:His parents require much of him.他父母对他要求甚高。本句可译为:特定情况所用材料的选择,在多数国家都取决于对公路服务质量的要求。

UNIT 15

Text A Bituminous Materials

According to nomenclature commonly in use in the United States, the term "bituminous material" is used to denote substances in which bitumen is present or from which it can be derived. Bitumen is a hydrocarbon material of either natural or pyrogenous origin, gaseous, liquid, semisolid, or solid, which is completely soluble in carbon disulfide (ASTM[①]).

With respect to the use in highway construction, the term bituminous material is used to include both natural and manufactured materials regardless of origin, but is restricted to those hydrocarbon materials which are cementitious in character or from which a residuum of this character will develop.

Terms Relating to Asphalt

Asphalt Black to dark brown, semisolid to solid cementitious materials consisting principally of bitumen that gradually liquefy when heated and which occur in nature as such[②] or are obtained as a residuums in the refining of petroleum.

Asphalt Cement or *Paving Asphalt* An asphalt specially prepared as to quality and consistency for direct use in paving. It has a normal penetration between 40 and 300 and must be used hot.

Native or *Natural Asphalt* One occurring as such in nature. It may be of the lake, rock, or vein variety, may be essentially pure bitumen or contain a large amount of mineral matter, and the asphalt may vary from hard to soft.

Petroleum Asphalt Asphalt produced by the refining of petroleum.

Cutback Asphalt Asphalt dissolved in naphtha (gasoline) or kerosene to render it temporarily fluid for use. If the solvent is kerosene and the dissolved asphalt is relatively soft, the cutback is designated as a medium-curing one.

Emulsified Asphalt A mixture in which an asphalt cement in a finely dispersed state is suspended in chemically treated water. An inverted asphalt emulsion is one in which asphalt is the continuous phase with water dispersed in it. Inverted asphalt emulsions used in road construction usually are made from liquid asphaltic materials.

Asphalt Concrete A plant mix of closely graded mineral aggregate and asphalt, designed and controlled to produce a mixture of high quality from the standpoint of both stability and durability. Such mixtures are usually produced hot with asphalt cement, but other types may be used as long as the resulting mixture is as described[③]. It may be produced as base, binder, or surface courses.

Words and Expressions

cementitious　*a.* 有黏性的,黏结的
denote　*v.* 表示,意味着
hydrocarbon　*n.* 碳氢化合物,烃
pyrogenous　*a.* 火成的,发热的
gaseous　*a.* 气体的,气态的
soluble　*a.* 可溶的,可解决的
residuum ［复］residua　*n.* 残渣,残余
disulfide/disulphide　*n.* 二硫化物
asphalt　*n.* 沥青
asphaltic　*a.* 沥青的
consistency　*n.* 稠度,黏度
penetration　*n.* 针入度,贯入
vein　*n.* 矿脉;静脉
naphtha　*n.* 轻汽油,挥发油
liquefy　*v.* 液化
refine　*v.* 精制,精炼,提炼;使变纯
refinery　*n.* 炼制厂
refining process　精炼法
cutback　*n.* 轻制
cutback asphalt　轻制沥青
dissolve　*v.* 溶解;分散;解散
kerosene　*n.* 煤油,火油
render　*v.* 致使,使成为
solvent　*n.* 溶剂
designate　*v.* 标明,指明
medium　*a.* 中位的,中间的;*n.* ［复］media　中介物,介质;中间
medium-curing　中凝
emulsify　*v.* 乳化
disperse　*v.* 分散,弥散,扩散

suspend   *v.* 悬,吊;悬浮
invert   *v.* 反转,倒置,颠倒
inverted asphalt emulsion   沥青倒乳液
durability   *n.* 耐久性,经久性
regardless of   无论,不管

Notes

①ASTM:American Society for Testing and Materials 美国试验与材料学会。
②as such 如此,像这样,such 是代词,作介词 as 的宾语。再如下文的 One occurring as such in nature.
③as described 是 as is described 的省略形式,作表语。

Text B   Bituminous Surface

Bituminous materials combined aggregates are the most common pavement surfaces in use today. They are used on all types of roadway from multiple layers of asphalt concrete on the highest class of highway to thin, dust-control layers on seldom-used roads.

Preliminary Treatments

In the construction of many types of bituminous pavements, a prime coat or tack coat is employed as part of the construction procedure.

Prime Coat   Liquid bituminous material applied to penetrate the surface of a base on which a bituminous pavement is to be placed. The prime coat is applied to bond together loose particles on the surface, to protect the surface from weather, traffic, and construction equipment, and to prepare the surface to receive the bituminous pavement.

Tack Coat   A bituminous material applied to provide bond between a new or old surface and the mixture with which it is being covered. It may be used in a addition to a prime coat and may be applied as a hot asphalt cement or as a liquid bituminous material.

Bituminous Surface Treatment

A bituminous surface treatment is generally constructed by making an application of bituminous material, immediately followed by the application of covering mineral aggregate. The surface is then rolled. If a single application of bituminous material is made, followed by a single application of aggregate, it may be called a single surface treatment. The thickness of this type of surface may be increased by the application of additional bituminous material and additional spreads of successively finer aggregate. The result is a double or triple surface treatment.

Bituminous surface treatments are used to waterproof the surface in order to prevent softening of the base and subgrade, and to prevent abrasion, raveling, dusting, and disintegration of the surface by traffic or propeller blast. They are especially adapted for surfacing worn or aged bituminous pavements, and on strong bases as a wearing surface for light loads.

Seal Coat

A special type of single surface treatment is a seal coat, which is used as a final layer in the construction of many bituminous pavements. Seal coats should never be used on bituminous concrete airfield pavements which will be subjected to jet aircraft traffic.

Penetration Macadam

Penetration macadam pavement differs from all other bituminous pavements in that the bitumen is added to the aggregate after the latter is in place on the base. Coarse, angular, crushed rock is spread and rolled, sprayed with a penetrating coat of bituminous materials, and covered with a smaller stone chip rolled and broomed to fill the voids in the surface. Penetration macadam is well adapted for use as a base course.

Mixed-in-place Bituminous Pavements (Road Mix)

Mixed-in-place bituminous pavements are constructed by mixing bitumen with aggregate directly on the road or runway. Mixing may be accomplished by the use of ordinary construction equipment or by traveling mixers. Mixed-in-place pavements provide an economical means of obtaining a satisfactory surface for roads when the required amount of pavement is small and when suitable natural aggregates are on or near the side. They are not considered suitable for permanent type pavement, but may be as a wearing course for temporary roads or as the first step in stage construction for higher type roads.

Hot-Mix Bituminous Concrete Pavements (Plant Mix)

Bituminous concrete is an intimate mixture of concrete of coarse aggregate, fine aggregate, mineral filler, and bituminous material, proportioned and mixed at a central mixing plant. Hot-mix bituminous concrete is mixed, spread, and compacted at elevated temperature. Coarse aggregate is material retained on the No.10 sieve; fine aggregate is that passing the No.10 sieve and retained on the No.200; and mineral filler is material finer than No.200 sieve. Bituminous concrete contains more than 35 to 40 percent coarse aggregate.

Sheet Asphalt

A hot plant mixture of sand, filler, and asphalt cement carefully proportioned and controlled to produce a sand-type mixture of high quality. It is usually used for surfacing courses.

Words and Expressions

aggregate　　*n.* 集料,粒料,掺和料
preliminary　　*a.* 预先的
prime coat　　透层
tack coat　　黏层
mineral　　*a.* 矿物的
roll　　*v.* 碾压
triple　　*a.* 三倍的
waterproof　　*a.* 防水的
abrasion　　*n.* 磨耗
ravel　　*v.* 使松散
disintegration　　*n.* 溃散
wear(wore,worn)　　*v.* 磨耗
seal coat　　封层
be subject to　　易受……的,常遭……的
angular　　*a.* 有棱角的
spray　　*v.* 喷射,喷涂,喷雾
void　　*n.* 空隙
runway　　*n.* 机场的跑道
chip　　*n.* 石屑
bituminous binder course　　沥青联结层
bituminous surface treatment　　沥青表面处治
penetration macadam　　灌入式碎石(沥青灌入式)
rapid-setting　　快裂
mixed-in-place(road) mix　　路拌混合料
blade grader　　平地机
travel-plant mix　　移动式设备拌和
plant mix　　厂拌(混合料)
cold-laid mixture　　冷铺混合料
sheet asphalt　　片沥青(砂粒式沥青混凝土)

UNIT 16

Text A Building Materials

Materials for building must have certain physical properties to be structurally useful. Primarily, they must be able to carry a load or weight, without changing shape permanently. When a load is applied to a structure member, it will deform: that is a wire will stretch or a beam will bend. However, when the load is removed, the wire and the beam come back to the original positions. This material property is called elasticity. If a material were not elastic and a deformation were present in the structure after removal of the load, repeated loading and unloading eventually would increase the deformation to the point where the structure would become useless. Materials used in architectural structures, such as stone and brick, wood, steel, aluminum, reinforced concrete, and plastics, behave elastically within a certain defined range of loading. If the loading is increased above the range, two types of behavior can occur: brittle and plastic. In the former, the material will break suddenly. In the latter, the material begins to flow at a certain load (yield strength), ultimately leading to fracture. As examples, steel exhibits plastic behavior, and stone is brittle. The ultimate strength of a material is measured by the stress at which failure (fracture) occurs.

A second important property of a building material is its stiffness. This property is defined by the elastic modulus, which is the ratio of the stress (force per unit area), to the strain (deformation per unit length). The elastic modulus, therefore, is a measure of the resistance of a material to deformation under load. For two materials of equal area under the same load, the one with the higher elastic modulus has the smaller deformation. Structural steel, which has an elastic modulus of 30 million pounds per square inch (psi), or 2,100,000 kilograms per square centimeter, is 3 times as stiff as aluminum, 10 times as stiff as concrete, and 15 times as stiff as wood.

Masonry

Masonry consists of natural materials, such as stone or manufactured products, such as brick and concrete blocks. Masonry has been used since ancient times; mud bricks were used in the city of Babylon for secular buildings, and stone was used for the great temples of the Nile Valley. The Great Pyramid in Egypt, standing 481 feet (147 meters) high, is the most spectacular masonry construction. Masonry units originally were stacked without using any bonding agent, but all modern masonry construction uses a cement mortar as a bonding material. Modern structural materials include stone, brick of burnt clay or slate, and concrete blocks.

Masonry is essentially a compressive material: it cannot withstand a tensile force, that is, a pull. The ultimate compressive strength of bonded masonry depends on the strength of the masonry unit and the mortar. The ultimate strength will vary from 1,000 to 4,000 psi (70 to 280 kg/cm^2), depending on the particular combination of masonry unit and mortar used.

Timber

Timber is one of the earliest construction materials and one of the few natural materials with good tensile properties. Hundreds of different species of wood are found throughout the world, and each species exhibits different physical characteristics. Only a few species are used structurally as framing members in building construction. In the United States, for instance, out of more than 600 species of wood, only 20 species are used structurally. These are generally the conifers, or softwoods, both because of their abundance and the ease with which their wood can be shaped. The species of timber more commonly used in the United States for construction are Douglas fir, Southern pine, spruce, and redwood. The ultimate tensile strength of these species varies from 5,000 to 8,000 psi (350 to 560 kg/cm^2). Hardwoods are used primarily for cabinetwork and for interior finishes such as floors.

Because of the cellular nature of wood, it is stronger along the grain than across the grain. Wood is particularly strong in tension and compression parallel to the grain. And it has great bending strength. These properties make it ideally suited for columns and beams in structures. Wood is not effectively used as a tensile member in a truss, however, because the tensile strength of a truss member depends upon connections between members. It is difficult to devise connections which do not depend on the shear or tearing strength along the grain, although numerous metal connectors have been produced to utilize the tensile strength of timbers[①].

Steel

Steel is an outstanding structural material. It has a high strength on a pound-for-pound basis when compared to other materials, even though its volume-for-volume weight is more than ten times that of wood. It has a high elastic modulus, which results in small deformations under load. It can be formed by rolling into various structural shapes such as I-beams, plates, and sheets; it also can be cast into complex shapes; and it is also produced in the form of wire strands and ropes for use as cables in suspension bridges and suspended roofs, as elevator ropes, and as wires for prestressing concrete. Steel elements can be joined together by various means, such as bolting, riveting, or welding. Carbon steels are subject to corrosion through oxidation and must be protected from contact with the atmosphere by painting them or embedding them in concrete. Above temperatures of about 700 F (371 ℃), steel rapidly loses its strength, and therefore it must be covered in a jacket of a fireproof material (usually concrete) to increase its fire resistance.

The addition of alloying elements, such as silicon or manganese, results in higher strength steels with tensile strengths up to 250,000 psi (17,500 kg/cm^2)[②]. These steels are used where the size of a structural member becomes critical, such as in the case of columns in a skyscraper.

Aluminum

Aluminum is especially useful as a building material when lightweight, strength, and corrosion resistance are all important factors. Because pure aluminum is extremely soft and ductile, alloying elements, such as magnesium, silicon, zinc, and copper, must be added to it to improve the strength required for structural use. Structural aluminum alloys behave elastically. They have an elastic modulus one third as great as steel and therefore deform three times as much as steel under the same load. The unit weight of an aluminum alloy is one third that of steel, and therefore an aluminum member will be lighter than a steel member of comparable strength. The ultimate tensile strength of aluminum alloys ranges from 20,000 to 60,000 psi (1,400 to 4,200 kg/cm^2).

Aluminum can be formed into a variety of shapes; it can be extruded to form I-beams, drawn to form wire and rods, and rolled to form foil and plates. Aluminum members can be put together in the same way as steel by riveting, bolting, and (to a lesser extent) by welding. Apart from its use for framing members in buildings and prefabricated housing, aluminum also finds extensive use for window frames and for the skin of the building in curtain-wall construction.

Concrete

Concrete is a mixture of water, sand and gravel, and portland cement. Crushed stone, manufactured lightweight stone, and seashells are often used in lieu of mural gravel. Portland cement, which is a mixture of materials containing calcium and clay, is heated in a kiln and then pulverized. Concrete derives its strength from the fact that pulverized portland cement, when mixed with water, hardens by a process called hydration. In an ideal mixture, concrete consists of about three fourths sand and gravel (aggregate) by volume and one fourth cement paste. The physical properties of concrete are highly sensitive to variations in the mixture of the components, so a particular combination of these ingredients must be custom-designed to achieve specified results in terms of strength or shrinkage. When concrete is poured into a mold or form, it contains free water, not required for hydration, which evaporates. As the concrete hardens, it releases this excess water over a period of time and shrinks. As a result of this shrinkage, fine cracks often develop. In order to minimize these shrinkage cracks, concrete must be hardened by keeping it moist for at least 5 days. The strength of concrete increases in time because the hydration process continues for years; as a practical matter, the strength at 28 days is considered standard.

Concrete deforms under load in an elastic manner. Although its elastic modulus is one tenth that of steel, similar deformations will result since its strength is also about one tenth that of steel. Concrete is basically a compressive material and has negligible tensile strength.

Reinforced Concrete

Reinforced concrete has steel bars that are placed in a concrete member to carry tensile forces. These reinforcing bars, which range in diameter from 0.25 inch (0.64 cm) to 2.25 inches (5.7 cm), have wrinkles on the surfaces to ensure a bond with the concrete. Although reinforced concrete was developed in many countries, its discovery usually is attributed to Joseph Monnier, a French gardener, who used a wire network to reinforce concrete tubes in 1868. This process is workable because steel and concrete expand and contract equally when the temperature changes. If this were not the case, the bond between the steel and concrete would be broken by a change in temperature since the two materials would respond differently. Reinforced concrete can be molded into innumerable shapes, such as beams, columns, slabs, and arches, and is therefore easily adapted to a particular form of building[③]. Reinforced concrete with ultimate tensile strengths in excess of 10,000 psi (700 kg/cm^2) is possible, although most commercial concrete is produced with strengths under 6,000 psi (420 kg/cm^2).

Plastics

Plastics are rapidly becoming important construction materials because of the great variety, strength, durability, and lightness. A plastic is a synthetic material or resin which can be molded into any desired shape and which uses an organic substance as a binder. Organic plastics are divided into two general groups: thermosetting and thermoplastic. The thermosetting group becomes rigid through a chemical change that occurs when heat is applied; once set, these plastics cannot be remolded. The thermoplastic group remains soft at high temperatures and must be cooled before becoming rigid; this group is not used generally as a structural material. The ultimate strength of most plastic materials is from 7,000 to 12,000 psi (490 to 840 kg/cm^2), although nylon has a tensile strength up to 60,900 psi (4,200 kg/cm^2).

Words and Expressions

    secular　　*a.* 世俗的,现世的,非宗教的
    temple　　*n.* 庙,寺,神殿,教堂
    pyramid　　*n.* 金字塔,四面体
    stack　　*v.* 堆叠,成堆,整齐地堆起
    masonry　　*n.* 石工,圬工;砖石建筑,砌筑
    masonry units　　砌块,砌体砌筑单元
    masonry construction　　圬工建筑,圬工工程,砖石工程,砖石结构

mortar　*n.* 砂浆
slate　*n.* 板岩,石板瓦,板石
timber　*n.* 木材,木料,原木
conifer　*n.* 针叶树,松柏类植物
douglas fir　花旗松
abundance　*n.* 丰富,充足,富裕,多
spruce　*n.* 云杉,云杉木
cabinetwork　*n.* 细木工,细木家具
cellular　*a.* 细胞的,分格的,多孔状的
grain　*n.* 颗粒,纹理,粒面
manganese　*n.* 锰
silicon　*n.* 硅
seashell　*n.* 贝壳,海贝
negligible　*a.* 可以忽略的,微不足道的
synthetic　*a.* 合成的,人造的,综合的
lieu　*n.* 场所
in lieu of　代,代替
resin　*n.* 树脂,胶质,人造树脂
thermosetting　*a.* 热凝性的,热固性的
truss　*n.* 桁(构)架,桁梁,构架工程
elevator ropes　(电梯)升运机缆索
ultimate tensile strength　极限抗拉强度
carbon steel　碳素钢
unit weight　密度
prefabricated housing　活动(预制)房屋
framing member　框架构件,构架件
window frame　窗框
in terms of…　在……方面

Notes

①It is difficult to devise connections which do not depend on the shear or tearing strength along the grain, although numerous metal connectors have been produced to utilize the tensile strength of timbers. 本句可译为:尽管可以利用木材的抗拉强度制造出若干金属节点,但很难设计出与顺纹剪切强度或抗裂强度关系不大的接头。

②The addition of alloying elements, such as silicon or manganese, results in higher strength steels with tensile strengths up to 250,000 psi. 本句可译为:合金元素(如硅或锰)的掺入使钢材强度变得更高,其抗拉强度可达 250,000 lb/in^2(17,500 kg/cm^2)。

③Reinforced concrete can be molded into innumerable shapes, such as beams, columns, slabs, and arches, and is therefore easily adapted to a particular form of building. 本句可译为:钢筋混凝土可被浇铸成各种形状,例如梁、柱、板和拱等,因此用它建造特定形状的建筑并不困难。

Text B Testing of Materials

The most common test of building materials is the strength test to destruction. This is partly because strength is a very important property of a building material, even a material in a "non-load-bearing" part of the building; partly because strength tests are comparatively simple to carry out; and partly because they offer a guide to other properties, such as durability.

The strength of a ductile material such as steel, aluminum, or plastics is usually determined by applying a tensile load. A compression test is used for brittle materials such as concrete, stone, and brick because their tensile strength is low and thus harder to measure accurately.

The method of testing and the dimensions of the test pieces are laid down in the appropriate standards published by the American Society for Testing Materials (ASTM), the British Standards Institution (BSI), the Standards Association of Australia (SAA), etc.

The size and shape of the test specimen are particularly important for brittle materials because they influence the number of flaws that are likely to occur in the test specimen. For concrete tests, the standard American and Australian test specimen—a cylinder 150 mm in diameter and 300 mm long—gives a lower result than the standard British test specimen—a 150 mm cube-because the former contains more concrete.

The speed of testing is also specified. A passage of time is required for both plastic deformation and the formation of cracks, and a faster rate of testing thus gives a higher result.

For tests on concrete and timber, it is necessary to specify the moisture content because this affects the strength.

A test on a single specimen is unreliable because we do not know whether it is an average test specimen or whether it has fewer or more than the average number of minute flaws. Standard specifications lay down how many specimens shall be tested and how they are to be selected.

Tests of factory-made materials carried out by the manufacturer are usually accepted by the user of the building material unless he has reason to doubt their veracity. Since concrete is made on the building site or brought from a ready-mix concrete plant, its testing becomes the responsibility of the building contractor. This is therefore a more frequent testing activity than that for other materials. Concrete cylinders or cubes are normally tested in a hydraulic press, which may be used exclusively for this purpose. A universal machine based on the same principle.

Timber differs from other building materials in that it is produced from growing trees and is thus more variable. Cut timber from virgin forest may consist of a variety of different species. Even timber cut from a planted forest containing trees of the same species all planted at the same time may show appreciable variation between pieces because of knots, or other flaws.

A substantial proportion of timber is used on domestic construction where it is not highly stressed; in such cases, "visual grading" (that is, merely looking at it) may be sufficient. Because of the imperfections in individual pieces, "stress grading" is usually more reliable than even accurate testing of selected test pieces. A stress grading machine tests every individual piece of timber by a method that is very fast and relatively cheap. The machine is based on an empirical relation between the strength and the deflection of timber. Each piece of timber is deflected (but not stressed to its limit) at several points along its length, and the deflection category marked by means of a spot of dye. The timber is then classified visually by its color markings.

The strength of metals is reduced if they are repeatedly loaded alternately in tension and in compression. This is called repeated loading if it is applied several hundreds or thousands of times, and fatigue loading if it is applied millions of times. Fatigue loading is a major problem in machines but rarely in buildings. Wind loads, however, can cause repeated loading in roof structures. There are special machines for testing the strength of materials under repeated loading.

Other special tests are for ductility and for hardness. Ductility is tested by bending a bar around a pin over a wide angle. Hardness is tested by indentation with a diamond or a hardened steel ball. The hardness test is carried out only if an accurate result is required because there is a good correlation between the tensile strength test and the various hardness tests for the metals. If the tensile strength has been tested, then the hardness of the metal can be deduced from that with reasonable tolerance.

The toughness of a metal can also be deduced from the tension test. Toughness is defined as the energy required to break a material. Energy is force multiplied by distance, that is, the integral of force in relation to length, or the area contained under a force-deformation curve. Stress is force per unit area, and strain is deformation per unit length, so that the area contained under the stress-strain diagram represents the energy per unit volume. The greater the area contained under a stress-strain curve up to failure, the greater the toughness of the material. Consequently, ductile materials that deform plastically are much tougher than brittle materials that show little plastic deformation.

Words and Expressions

destruction   *n.* 破坏,破裂,毁坏
non-load-bearing   非承重,不承重
durability   *n.* 耐久性,持久性,使用年限
brittle   *a.* 脆的,脆性的
BSI   英国标准协会
SAA   澳大利亚标准协会
flaw   *n.* 缺陷
cylinder   *n.* 圆柱体
passage   *n.* 通过,行程,一段,一节
timber   *n.* 木材,木料
moisture content   含水量

minute   *a.* 微小的,细微的,精密的,细致的
veracity   *n.* 诚实,真实性,准确性,精确性
ready-mixed concrete   预拌混凝土
building contractor   建筑承包商
hydraulic press   水压机
virgin forest   原始森林
a variety of   大量的
appreciable   *a.* 看得出的,显著的
domestic   *a.* 房屋的
grading   *n.* 分等,分级,分类,级配
reliable   *a.* 可靠的
fatigue   *n.* 疲劳
correlation   *n.* 相关关系
tolerance   *n.* 公差
toughness   *n.* 韧性,韧度
multiply   *v.* 乘
stress-strain curve   应力-应变曲线
deduce   *v.* 推断,推定,导出,演绎

UNIT 17

Text A Stress-Strain Relationship of Materials

The satisfactory performance of a structure frequently is determined by the amount of deformation or distortion that can be permitted. A deflection of a few thousandths of an inch might make a boring machine useless, whereas the boom on a dragline might deflect several inches without impairing its usefulness. It is often necessary to relate the loads on a structure, or on a member in a structure, to the deflection the loads will produce. Such information can be obtained by plotting diagrams showing loads and deflections for each member and type of loading in a structure, but such diagrams will vary with the dimensions of the members, and it would be necessary to draw new diagrams each time the dimensions were varied. A more useful diagram is one showing the relation between the stress and strain. Such diagrams are called stress-strain diagrams.

Data for stress-strain diagrams are usually obtained by applying an axial load to a test specimen and measuring the load and deformation simultaneously. A testing machine is used to strain the specimen and to measure the load required to produce the strain. The stress is obtained by dividing the load by the initial cross-sectional area of the specimen. The area will change somewhat during the loading, and the stress obtained using the initial area is obviously not the exact stress occurring at higher loads. It is the stress most commonly used, however, in designing structures. The stress obtained by dividing the load by the actual area is frequently called the true stress and is useful in explaining the fundamental behavior of materials. Strains are usually relatively small in materials used in engineering structures, often less than 0.001, and their accurate determination requires special measuring equipment.

True strain, like true stress, is computed on the basis of the actual length of the test specimen during the test and is used primarily to study the fundamental properties of materials. The difference between nominal stress and strain, computed from initial dimensions of the specimen, and true stress and strain is negligible for stresses usually encountered in engineering structures, but sometimes the difference becomes important with larger stresses and strains.

The initial portion of the stress-strain diagram for most materials used in engineering structures is a straight line. The stress-strain diagrams for some materials, such as gray cast iron and concrete, show a slight curve even at very small stresses, but it is common practice to draw a straight line to average the data for the first part of the diagram and neglect the curvature. Thomas Young, in 1807, suggested using the ratio of stress to strain to measure the stiffness of a material. This ratio is called Young's modulus or the modulus of elasticity and is the slope of the straight line portion of the stress-strain diagram. Young's modulus is written as

$$E = \sigma/\varepsilon \quad \text{or} \quad G = \tau/\gamma$$

where E is used for normal stress and strain and G (sometimes called the modulus of rigidity) is used for shearing stress and strain. The maximum stress for which stress and strain are proportional is called the proportional limit.

The action is said to be elastic if the strain resulting from loading disappears when the load is removed. The elastic limit is the maximum stress for which the material acts elastically.

When the stress exceeds the elastic limit (or proportional limit for practical purposes), it is found that a portion of the deformation remains after the load is removed. The deformation remaining after an applied load is removed is called plastic deformation. Plastic deformation independent of the time duration of the applied load is known as slip. Creep is plastic deformation that continues to increase under a constant stress. In many instances creep continues until fracture occurs; however, in other instances the rate of creep decreases and approaches zero as a limit. Some materials are much more susceptible to creep than others, but most materials used in engineering exhibit creep at elevated temperatures. The total strain is thus made up of elastic strain, and possibly combined with plastic strain that results from slip, creep, or both. When the load is removed, the elastic portion of the strain is recovered, but the plastic part (slip and creep) remains as permanent set.

A precise value for the proportional limit is difficult to obtain, particularly when the transition of the stress-strain diagram from a straight line to a curve is gradual. For this reason, other measures of stress that can be yield strength for a specified offset are frequently used for this purpose.

The yield point is the stress at which there is an appreciable increase in strain with no increase in stress, with the limitation that, if straining is continued, the stress will again increase.

The yield strength is defined as the stress that will induce a specified permanent set, usually 0.05 to 0.3 percent, which is equivalent to a strain of 0.000 5 to 0.003. The yield strength is particularly useful for materials with no yield point.

The maximum stress, based on the original area, developed in a material before rupture is called the ultimate strength of the material, and the term may be modified as the ultimate tensile, compressive, or shearing strength of the material. Ductile materials undergo considerable plastic tensile or shearing deformation before rupture. When the ultimate strength of a ductile material is reached, the cross-sectional area of the test specimen starts to decrease or neck down, and the resultant load that can be carried by the specimen decreases. Thus, the stress based on the original area decreases beyond the ultimate strength of the material, although the true stress continues to increase until rupture.

Words and Expressions

distortion   *n.* 变形,挠曲,扭曲,歪曲
boring machine   镗床
boom   *n.* 吊杆,起重杆,悬臂
dragline   *n.* 拉铲挖土机,挖掘斗
impair   *v.* 削弱,损害
plot   *v.* 绘制,标绘
specimen   *n.* 试件
nominal   *a.* 名义上的,有名无实的
negligible   *a.* 可以忽视的,微不足道的
gray cast iron   灰铸铁
ratio   *n.* 比值
rigidity   *n.* 刚性,刚度,稳定性
proportional limit   比例极限
elastic limit   弹性极限
plastic deformation   塑性变形
slip   *n.* 滑动,滑移,打滑
creep   *n.* 爬行,蠕变,徐变
fracture   *n.* 折断,断裂
yield strength   *n.* 屈服强度
rupture   *n.* 破裂,断裂,破坏
modify   *v.* 修改
ductile   *a.* 可延展的
neck down   颈缩

Text B   Load Classification

The primary objective of a course in mechanics of materials is the development of relationships between the loads applied to a non-rigid body and the internal forces and deformations induced in the body. Ever since the time of Galileo Galilei (1564—1642), men of scientific bent have studied the problem of the load-carrying capacity of structural members and machine components, and have developed mathematical and experimental methods of analysis for determining the internal forces and the deformation induced by the applied loads. The experiences and observations of these scientists and engineers of the last three centuries are the heritage of the engineer of today. The fundamental knowledge gained over the last three centuries, together with the theories and analysis techniques developed, permit the modern engineer to design, with complete competence and assurance, structures and machines of unprecedented size and complexity.

It will frequently be found that the equations of equilibrium (or motion) are not sufficient to determine all the unknown loads or reactions acting on a body. In such cases it is necessary to consider the geometry (the change in size or shape) of the body after the loads are applied. The deformation per unit length in any direction or dimension is called strain. In some instances, the specified maximum deformation and not the specified maximum stress will govern the maximum load that a member may carry.

Certain terms are commonly used to describe applied loads; their definitions are given here so that the terminology will be clearly understood.

Loads may be classified with respect to time:

(1) A static load is a gradually applied load for which equilibrium is reached in a relatively short time.

(2) A sustained load is a load that is constant over a long period of time, such as the weight of a structure (called dead load). This type of load is treated in the same manner as a static load; however, for some materials and conditions of temperature and stress, the resistance to failure may be different under short-time loading and under sustained loading.

(3) An impact load is a rapidly applied load (an energy load). Vibration normally results from an impact load, and equilibrium is not established until the vibration is eliminated, usually by natural damping forces.

(4) A repeated load is a load that is applied and removed many thousands of times. The helical springs that close the valves on automobile engines are subjected to repeated loading.

Loads may also be classified with respect to the area over which the load is applied:

(1) A concentrated load is a load or force applied at a point. Any load applied to a relatively small area compared with the size of the loaded member is assumed to be a concentrated load; for example, a truck wheel load on the longitudinal members of a bridge.

(2) A distributed load is a load distributed along a length or over an area. The distribution may be uniform or nonuniform. The weight of a concrete bridge floor of uniform thickness is an example of a uniformly distributed load.

Loads may be classified with respect to the location and method of application:

(1) A centric load is one in which the resultant force passes through the centroid of the resisting section. If the resultant passes through the centroids of all resisting sections, the loading is termed axial.

(2) A torsional load is one that subjects a shaft or some other member of couples that twist the member. If the couples lie in planes transverse to the axis of the member, the member is subjected to pure torsion.

(3) A bending or flexural load is one in which the loads are applied transversely to the longitudinal axis of the member. A member subjected to bending loads bends or bows along its

length.

(4) A combined loading is a combination of two or more of the previously defined types of loading.

Words and Expressions

non-rigid   *a.* 非刚性的
heritage   *n.* 遗产
assurance   *n.* 保证,把握,自信
unprecedented   *a.* 无先例的,史无前例的,崭新的
equation   *n.* 公式
equilibrium   *n.* 平衡
geometry   *n.* 几何学
terminology   *n.* 专用名词,术语,词汇
static load   *n.* 静荷载
sustained load   *n.* 持续荷载
impact load   *n.* 冲击荷载
repeated load   *n.* 重复荷载
helical   *n. & a.* 螺旋面,螺线,螺旋状的
helical spring   螺旋形弹簧
classify   *v.* 分类
concentrated load   *n.* 集中荷载
longitudinal   *a.* 经度的
distributed load   *n.* 分布荷载
uniform   *a.* 均匀的,统一的
resultant   *a.* 合成的,综合的,总的; *n.* 合力
centroid   *n.* 矩心,质心,重心,形心
resisting section   反力面
axial   *a.* 轴向的
torsional   *a.* 扭的,转的
bending   *n.* 挠曲,弯曲,扭弯
flexural   *a.* 弯曲的,挠性的
couple   *n.* 力偶,力矩

UNIT 18

Text A Field Measurement of Density and Moisture Content

Control of field compaction requires frequent determination of the density and moisture content of the compacted soil, and also measurements of the moisture content of the excavated soil to assess its suitability as filling material. There is a number of standard methods of measuring field dry density, the two most commonly used being the sand-replacement method and the core-cutter method[①]. Both these methods are outlined briefly below; in each case the bulk density of the soil is determined by finding the weight of material occupying a known volume and, by measuring the moisture content of a representative sample, the dry density can then be calculated.

Sand-replacement Method

A hole approximately 4 in. diameter (8 in. diameter for coarse-grained soils) is excavated to the depth of the layer to be tested and the material removed is weighed. The volume of the hole is then determined by running in dry, closely-graded sand of known bulk density from a suitable container, the volume of sand used having been found by weighing the sand-filled container before and after the test. The test may be performed on all types of soils, but is liable to give too high a value for the bulk density of wet granular materials as a result of the surrounding soil flowing into the sample hole during excavation.

Core-cutter Method

This method is useful for soft, cohesive soils which are free from stones. A 4-in. internal diameter cutter of known weight and volume is driven into the soil and then dug out. The soil sample is trimmed flush with the ends of the cutter and weighed to determined the bulk density.

With any of the standard methods for dry density, it is necessary to make a number of measurements over a comparatively small area in order to obtain a representative average value, since considerable variations occur as a result of testing errors. It has been suggested that between 5 and 10 measurements per 1,000 cubic yard of fill constitute a reasonable number as regards the amount of testing time involved, but recent work indicates that about 40 dry density and moisture content measurements per 1,000 cubic yard of fill are necessary to obtain a mean value with an accuracy not poorer than ±2 for the percentage air voids at the 90% confidence limits[②]. Clearly this number of tests would be impracticable using the standard methods of measurement, but efforts are at present being made to develop methods which can be carried out with greater rapidity.

One possible approach is based on the fact that the scattering and absorption of gamma radiation is a function of the bulk density of the material. Briefly, a radioactive source (caesium 137) contained in the end of a stainless-steel probe is inserted in the soil to a depth of up to 6 in. and a Geiger-Muller tube which detects gamma radiation is placed on the surface about 8 in. from the probe to measure the rate of transmission of radiation through the material. Alternatively, in soils containing large stones which make it difficult to insert the probe without causing undue disturbance, both the source and detector may be placed on the surface separated by lead shielding and the back-scatter method used, although this method suffers from the disadvantage that measurements are largely confined to the upper 2 in. of soil.

Since this equipment records the bulk density, it is necessary to determine also the moisture content of the material before the dry density can be calculated. This may also be done with nuclear techniques using a radium/beryllium source of fast neutrons placed on the ground surface alongside a boron trifluoride counter which records the number of slow neutrons scattered back after collisions have occurred between the fast neutrons and nuclei of hydrogen present in the soil moisture. This measurement is also made on a unit volume basis and therefore the difference of the two readings give the dry density of the soil.

Words and Expressions

measurement   *n.* 测量
density   *n.* 密度
moisture content   含水量
compaction   *n.* 压实,夯实
access   *v.* 估计,评价
suitability   *n.* 适用性
sand-replacement method   换砂法
core-cutter method   钻心法
bulk density   毛体积密度
sample   *n.* 试样
run in   倒入,注入
liable   *a.* 易于……的,有……倾向的

granular   *a.* 粒状的,粒料的
trim   *v.* 修整,刨平
flush   *v.* & *a.* 使齐平;齐平的
flush with   与……齐平
as regards   关于,至于
confidence   *n.* 信任,信心
confidence limit   置信界限(用于数理统计)
scatter   *v.* 散射,分散
absorption   *n.* 吸收
gamma   *n.* 伽马(希腊字母)
radiation   *n.* 辐射,放射
radioactive   *a.* 放射性的
caesium   *n.* 铯(化学元素)
probe   *n.* 探头
disturbance   *n.* 扰动,妨碍
shield   *vt.* 用盾掩护;*n.* 防护
nuclear   *a.* 核的,原子核的
radium   *n.* 镭(化学元素)
beryllium   *n.* 铍(化学元素)
neutron   *n.* 中子
boron   *n.* 硼(化学元素)
trifluoride   *n.* 三氟化合物
collision   *n.* 碰撞
nucleus   *n.* [复] nuclei   核,原子核

Notes

①There is a number of standards of methods of measuring field dry density, the two most commonly used being the sand-replacement method and the core-cutter method. 本句可译为:野外量测干密度的标准方法有多种,其中最常用的两种是换砂法和钻芯法。在"There be"句型中,在系动词 be 之后紧跟着像 a number of, a lot of 这类词组来修饰作主语的复数名词,但由于不定冠词 a 紧跟在系动词之后,此处习惯用 is 而不用 are。

②It has been suggested that between 5 and 10 measurements per 1,000 cu yd of fill constitute a reasonable number as regards the amount of testing time involved, but recent work indicates that about 40 dry density and moisture content measurements per 1,000 cu yd of fill are necessary to obtain a mean value with an accuracy not poorer than±2 for the percentage air voids at the 90% confidence limits. 本句可译为:据建议,考虑测试所需的时间,每 1 000 立方码的填土测定6~10次作为一个适宜的次数,但近期研究表明,对于在 90% 置信界限时的空隙率来说,为了使干密度和含水量的测定获得精确度不低于±2 的平均值,每 1 000 立方码的填土需作大约 40 次测

定。between 5 and 10 是主语 measurements 的定语,这里是介词短语作定语用。

Text B Tests for Determining the Density of Soils

Tests for density may be divided into two classes: laboratory tests to set a standard for density, and field tests to measure the density of a soil in place in the roadway structure. Laboratory tests may in turn be subdivided on the basis of compaction procedure, into "static", "dynamic" or "impact" methods.

Static Tests

Some agencies have used a static test to determine maximum density of laboratory samples. One such test is conducted as follows. About 4,000 g of soil containing a designated percentage of water are placed in a cylindrical mold 6 in. in diameter and 8 in. high. The sample is compressed under a load of 2,000 lb/in., applied at a speed of 0.05 in./min. When the full load is reached, it is held for a period of 1 min. and then gradually released. Using the known dry weight of soil, mold diameter, and the measured height, the dry density of the sample is computed. Enough samples are processed to delineate the peak of the moisture—density curve. This peak value represents the standard.

Dynamic or Impact Tests

Most agencies determine optimum moisture content and maximum density with dynamic or impact tests. Samples of soil, each containing a designated percentage of water, are compacted in layers into molds of specified size. Compaction is obtained with a given number of blows from a free-falling hammer of prescribed dimension and weight that has a flat, circular face. The peak of the moisture-density curve represents standard density.

Field Tests for Density of Soils in Place

Field Density and Moisture Content by Sampling

The procedure for determining relative compaction by sampling is as follows:
1) Dig a small sample of the compacted material through the full depth of the layer to be tested.
2) Obtain the wet and dry weights of the sample. From these also determine the moisture content of the sample.
3) Determine the volume that the sample occupied in the fill by finding how many pounds of a material of known unit weight are required to fill this space. Sand or water poured or pumped into a

flexible rubber liner have served this purpose.

4) From the dry weight of the sample and the known volume that it occupies in the fill, obtain the dry weight per cubic foot.

5) Determine the relative compaction of the soil in the fill by dividing its dry weight per cubic foot by the laboratory standard density.

Field Density and Moisture Content Measurement with Nuclear Devices

In recent years, the state transportation agencies and large local agencies have increasingly adopted nuclear devices for determining in-place densities and moisture contents.

The principle of measurement by nuclear means is relatively simple. A source of nuclear energy provides both gamma and high-velocity neutrons. The gamma rays are reflected to a degree dependent of the density of the material through which they travel. The high-velocity neutrons are slowed down by any hydrogen atoms they contact, hence more neutrons are slowed down if a given soil has a higher moisture content. The intensity of the reflected gamma rays and slowed down neutrons are measured separately by Geiger-Muller counter tubes. Gauge readings are easily converted to density and percent moisture using calibration curves or microprocessors.

Nuclear devices overcome some of the disadvantages of the sampling method. Since determinations can be taken quickly and while construction equipment is operating, delays are minimized, and this can reduce construction costs. Also, more readings can be taken, which leads to greater confidence that prescribed densities have been obtained and also permits a statistical approach to compaction control.

Words and Expressions

field test　现场测试
dynamic　*a.* 动力的
designate　*v.* 标示,指明
cylindrical　*a.* 圆柱体的
release　*v.* 释放
delineate　*v.* 描出……的外形,描绘
optimum　*a.* 最佳的
prescribe　*v.* 规定
gauge　*n.* 表,量器
convert　*v.* 转换
calibration　*n.* 校准;标定
statistical　*a.* 统计的,统计学的

UNIT 19

Text A   America on Wheels

Early automobiles were sometimes only "horseless carriages" powered by gasoline or steam engines. Some of them were so noisy that cities often made laws forbidding their use because they frightened horses.

Many countries helped to develop the automobile. The internal-combustion engine was invented in Austria, and France was an early leader in automobile manufacturing. But it was in the United States after 1900 that the automobile was improved most rapidly. As a large and growing country, the United States needed cars and trucks to provide transportation in places not served by trains.

Two brilliant ideas made possible the mass production of automobiles. An American inventor named Eli Whitney thought of one them, which is known as "standardization of parts". In an effort to speed up production in his gun factory, Whitney decided that each part of a gun could be made by machines so that it would be exactly like all the others of its kind. For example, each trigger would be exactly like all other triggers. A broken trigger could then be replaced immediately by an identical one. After Whitney's idea was applied to automobile production, each part no longer had to be made by hand. Machines were developed that could produce hundreds, even thousands, of identical parts that would fit into place easily and quickly.

Another American, Henry Ford, developed the idea of the assembly line. Before Ford introduced the assembly line, each car was built by hand. Such a process was, of course, very slow. As a result, automobiles were so expensive that only rich people could afford them. Ford proposed a system in which each worker would have a special job to do. One person, for example, would make only a portion of the wheels. Another would place the wheels on the car. And still another would insert the bolts that held the wheels to the car. Each worker needed to learn only one or two routine tasks.

But the really important part of Ford's idea was to bring the work to the worker. An automobile frame, which looks like a steel skeleton, was put on a moving platform. As the frame moved past the workers, each worker could attach a single part. When the car reached the end of the line, it was completely assembled. Oil, gasoline, and water were added, and the car was ready to be driven away. With the increased production made possible by the assembly line, automobiles became much cheaper, and more and more people were able to afford them.

Today it can be said that wheels run America. The four rubber tires of the automobile move America through work and play. Wheels spin, and people drive off to their jobs. Tires turn, and people shop for the week's food at the big supermarket down the highway. Hubcaps whirl, and the whole family spends a day at the lake. Each year more wheels crowd the highways as 10 million new cars roll out of the factories. One out of every six Americans works at assembling cars, driving trucks, building roads, or pumping gas. America without cars? It's unthinkable.

But even though the majority of Americans would find it hard to imagine what life could be like without a car, some have begun to realize that the automobile is a mixed blessing. Traffic accidents are increasing steadily, and large cities are plagued by traffic congestion. Worst of all, perhaps, is the air pollution caused by the internal-combustion engine. Every car engine burns hundreds of gallons of fuel each year and pumps hundreds of pounds of carbon monoxide and other gases into the air. These gases are one source of the smog① that hangs over large cities. Some of these gases are poisonous and dangerous to health, especially for someone with a weak heart or respiratory disease.

One answer to the problem of air pollution is to build a car that does not pollute. That's what several major automobile manufacturers are trying to do. But building a clean car is easier said than done②. So far, progress has been slow. Another solution is to eliminate car fumes altogether by getting rid of the internal-combustion engine. Inventors are now working on turbine-powered cars, as well as on cars powered by steam and electricity. But most of us won't be driving cars run on batteries or boiling water for a while yet. Many auto makers believe that it will take years to develop practical models that are powered by electricity or steam.

To rid the world of pollution—pollution caused not just by cars, but by all of modern industrial life—many people believe we must make some fundamental changes in the way many of us live. Americans may, for example, have to cut down on the number of privately owned cars and depend more on public mass transit systems. Certainly the extensive use of new transit systems could cut down on traffic congestion and air pollution. But these changes do not come easily. Sometimes they clash head on with other urgent problems. For example, if a factory closes down because it cannot meet government pollution standards, a large number of workers suddenly find themselves without jobs. Questioning the quality of the air they breathe becomes less important than worrying about the next paycheck.

But drastic action must be taken if we are to reduce traffic accidents, traffic congestion, and air pollution. While wheels have brought better and more convenient transportation, they have also brought new and unforeseen problems. Progress, it turns out, has more than one face.

Words and Expressions

gasoline *n.* 汽油
internal-combustion engine 内燃机
brilliant *a.* 卓越的,英明的
standardization *n.* 标准化
trigger *n.* 扳机
identical *a.* 完全相同的
assembly line 装配线
bolt *n.* 螺栓
routine *a.* 常规的,机械的
platform *n.* 工作台
spin *v.* 旋转
plague *v.* 折磨,困扰,使受煎熬
carbon monoxide 一氧化碳
respiratory *a.* 呼吸的
fume *n.* 烟,汽
public mass transit system 公共运输系统(公交公司)
clash *v.* 冲突
paycheck *n.* 工资单
drastic *a.* 激烈的,极端的

Notes

①smog 即 smoke 与 fog 组合而成。
②是 a case of easier said than done 的省略形式。

Text B Bus Priorities

Buses are able to carry anything up to about 80 passengers and those running on local services within large towns frequently do, particularly in the peaks. This is in sharp contrast to the private car whose average occupancy is about 1.6 persons. Certainly, the bus is a larger vehicle and take up about four times the road space compared with the car, it can carry 50 times more passengers. When road space is at a premium, therefore, the need for moving people by public rather than private transport is very strong.

Bus priority measures may take the following forms:

(1) Facilities for stopping on freeways and other roads where parking, loading or unloading is

prohibited.

(2) Authority to make right turns (or left turns where the rule of the road is to keep to the right) barred to other traffic for the purpose of reducing conflicting vehicular movements.

(3) Activation of traffic light in their favor by buses by means of special equipment placed on the vehicle.

(4) Special bus lanes (usually the near side lane) which allows buses (in single file) to proceed ahead of other road users held in traffic blocks. (The practicability of a bus lane is dependent of course on there being an adequate width of road to allow at least a second lane for general traffic).

(5) Contra-flow operation along what have otherwise become one-way streets. This is, in effect, an extension of the bus only lane principle but against the normal traffic flow.

(6) The use of through routes denied to other traffic by the provision of special "bus gates" being, in effect, no entry signs which buses are permitted to pass.

(7) The use of roads denied to all other traffic; in other words, bus only roads.

(8) Special provision for buses built into a system of urban traffic control. This could take the form of special computer programming for selected bus routes, separate and unrestricted access for buses on to a highway for which other traffic might have to queue and participation by bus operators in the control of such schemes.

Words and Expressions

priority   *n.* 优先权
peak   *n.* 高峰
in sharp contrast to   完全相反的
occupancy   *n.* 乘坐
at a premium   (喻)非常珍贵
bar   *v.* 阻拦,拦住
conflict   *v. & n.* 冲突
activation   *n.* 活化
traffic block   交通阻塞
contra-flow operation   逆流行驶

UNIT 20

Text A Traffic Engineering(I)

What is Traffic Engineering

Traffic Engineering is still a relatively new discipline within the overall bounds of civil engineering. It has nevertheless already been partially subsumed within the still newer but broader discipline of transportation planning. Transportation planning is concerned with "the planning, functional design, operation and management of facilities for any mode of transportation in order to provide for the safe, rapid, comfortable, convenient, economical and environmentally compatible movement of people and goods". Within that broad scope, traffic engineering deals with those functions in respect of roads, road networks, terminal points (i.e. parks), abutting lands and their relationships with other modes of transportation.

Those definitions, based on the 1976 ones of the U.S. Institute of Transportation Engineers, are compatible with, but in the light of changing public attitudes[①], are more complete than, the British Institution of Civil Engineers 1959 definition of traffic engineering, which is: "That part of engineering which deals with traffic planning and design of roads, of frontage development and of parking facilities and with the control of traffic to provide safe, convenient and economical movement of vehicles and pedestrians".

Traffic Problem

The problem is large and still growing. In 1920 the total number of motor vehicles licensed in Great Britain was 650,000. Fifty year later the comparable figure was 14,950,000—a growth factor of 23 times. In recent years the rate of growth has slackened somewhat, but it is still considerable.

In order to see the problem in everyday terms, consider High Street, anywhere. Assuming that present trends continue, it is expected that within the next fifteen years or so the traffic on this road will increase by around forty to fifty percent. If this increased volume of traffic were to be accommodated at the same standard as today, the road might need to be widened by a similar forty to fifty percent[2]—perhaps an extra lane of traffic for the pedestrian to cross. In many cases, to accommodate the foreseeable future demand would destroy the character of the whole urban environment, and is clearly unacceptable.

The traffic problem is of world-wide concern, but different countries are obviously at different stages in the traffic escalation—with America the lead. While a country has few roads and a relatively low standard of living there is little demand for motor transport—and no real traffic problem. As soon as the country is opened up by a road system, the standard of living and the demand for motor transport both rise, gathering momentum rapidly. Eventually—and the stage at which this happens is open to considerable debate—the demand for cars, buses and lorries become satiated. The stage is known as saturation level.

For comparison, car ownership figures in different countries in 1970 were:

India	0.01 cars/person
Israel	0.05 cars/person
Japan	0.09 cars/person
Ireland	0.13 cars/person
Netherlands	0.20 cars/person
Great Britain	0.21 cars/person
West Germany	0.23 cars/person
Australia	0.31 cars/person
USA	0.44 cars/person

But the growth in vehicle ownership is only part of the overall traffic problem. Obviously, if a country has unlimited roads of extreme width, the traffic problem would not rise. No country in the world could meet this requirement: apart from anything else, it would not make economic sense[3]. In Britain, in 1970 there were only 22 meters of (full width) road length for each vehicle using the roads[4]. This figure is decreasing steadily.

Three Possible Solutions

The basic problem of traffic is therefore simple—an ever-increasing number of vehicle seeking to use too little road space. The solution to the problem is also a not - too- difficult choice from three possibilities:

(1) Build sufficient roads of sufficient size to cope with the demand.

(2) Restrict the demand for roads by restricting the numbers of licensed vehicles.

(3) A compromise between (1) and (2)—build some extra roads, using them and the existing road network to their full potential, and at the same time apply some restraint measures, limiting the increase in demand as far as possible.

With no visible end to the demand yet in sight, and with modern road-making costs ranging around £1 million per kilometer, the cost of building roads in Britain to cope with an unrestricted demand would be far too great. Added to this, such colossal use of space in a crowded island cannot be seriously considered. In Los Angles, a city built around the motor-car, approximately half the land space is devoted to movement and parking space for the automobile. Our cities are already largely built—and no one would consider ruining their character by pulling them down and rebuilding around the car. Thus the first possible solution is ruled out.

Even today, in an age of at least semi-affluence in most of the Western World, the car is still to some extent a status symbol, a symbol of prestige. Every family wants to own one, and takes steps—saving or borrowing—to get one. As incomes and standards rise the second car becomes the target. Any move to restrict the acquisition of the private car would be most unpopular—and politically unlikely.

For many purposes the flexibility of the private car—conceptually affording door-to-door personal transport—is ideal, and its use can be accommodated. For the mass movement of people along specific corridors within a limited period of time—i.e. particularly the journey to work—it may be more efficient. Some sort of compromise solution is the inevitable answer to the basic traffic problem.

It is in the execution of the compromise solution that traffic engineering comes into its own[5]. Traffic engineering ensures that any new facilities are not over-designed and are correctly located to meet the demand. It ensures that the existing facilities are fully used, in the most efficient manner. The fulfillment of demand: making the use of the car less attractive in the peak periods in order that the limited road space can be more efficiently used by public transport. (A standard family car, carrying 1.8 persons, occupies over 3 m^2 per person compared with around 0.7 m^2 per person on a bus carrying perhaps 40 passengers.) Such restraint measures will often be accompanied by improvements in the public transport services, to accommodate the extra demand for them.

Words and Expressions

discipline   *n.* 学科
bound   *n.* 范围
nevertheless   *ad. & conj.*   然而,不过,仍然
subsume   *v.* 包括,把……列入
terminal point   终点
compatible   *a.* 相容的,协调的,一致的
in (the) light of   按照,根据,由于
license (=licence)   *n. & v.* 许可证;发许可证给……;批准
slacken   *v.* 变(放)慢
accommodate   *v.* 容纳,接纳
foreseeable   *a.* 可预见到的
urban   *a.* 城市的,都市的
escalation   *n.* (逐步)升级,逐步上升

momentum   *n.* 势头
satiate   *v.* 使满足,使吃饱
saturation   *n.* 饱和状态
cope with   适应,对付,妥善处理,解决,克服
compromise   *n.* 妥协,和解
colossal   *a.* 巨(庞)大的,非常的
around   *prep.* 根据,以……为基础
largely   *ad.* 基本上
pull down   摧(拆)毁
rule out   排除,取消,拒绝考虑
affluence   *n.* 富(足)裕
status   *n.* 情况,状况
symbol   *n.* 象征
prestige   *n.* 威信,声望(誉)
acquisition   *n.* 获得
flexibility   *n.* 灵活性
conceptually   *ad.* 概念地

Notes

①… but in the light of changing public attitudes… 为介词短语,作状语修饰 are more complete than…。

②If this increased volume of traffic were to be accommodated at the same standard as today, the road might need to be widened by a similar forty to fifty percent… 如果这一交通增长量要用今天的标准来容纳,则道路需要相应加宽 40%～50%。主句和从句中的动词均为虚拟语气。

③… apart from anything else, it would not make economic sense. 且不说别的什么(理由),经济上是没有意义的。

④In Britain, in 1970 there were only 22 meters of road length for each vehicle using the roads. 1970 年,在英国一辆汽车仅能分摊到 22 米的道路。

⑤It is in the execution of the compromise solution that traffic engineering comes into its own. 本句为强调句型,强调 in the execution of the compromise solution。

Text B   Traffic Engineering(Ⅱ)

Detail Aspects of the Problem

By derivation from the definitions of the traffic engineering, the traffic problem can be considered from several aspects, all of which are interconnected. The operative words in the definitions are "... safe ... convenient... economical... movement...". We can think of these words as embracing traffic flow, traffic speed, traffic safety, and amenities for traffic, with which aspects traffic economics is closely interwoven; and throughout, the concern for environmental compatibility[①].

Traffic has to flow along our roads and it can readily be appreciated that the actual volume of traffic that can move to and from will depend on the widths and alignments of these roads. Equally the traffic flow will depend on the speed of the traffic stream. The overall traffic speed is not just a matter of the speed of movement; it is also very much dependent on the amount of time for which the traffic delayed, largely at intersections. The definition calls for "safe... movement", which naturally leads us to consider traffic safely—how are accidents caused, where do they occur, how can we reduce them?

The terminal facility—the parking place—at the temporary end of a vehicle trip is as important as the road leading to it. Parking facilities must be used efficiently and if new parks are to be provided, their size and location need to be carefully considered. Equally, in coping with the traffic problem as a whole it is essential that the human being remain the master and not the slave of the motor-car. In directing traffic along existing roads the effect of this on the existing environment must be assessed, or we have a "non-amenity".

In order to get the most roads and road use for our money—assuming that is the best object on which to spend scarce transportation resources—the cost of every road project must be investigated. To assess real costs, which quantitatively of themselves mean little, the benefits derived by society from the project are also considered, and sometimes the rate of return from the capital investment is compared with other projects. Thus, a great saving in road-user costs may not be economically justifiable to build a new bypass at all. Economics comes into all aspects of traffic engineering.

Tackling the Problem

Having now considered in detail what the problem is, in all its aspects, we can tackle it. There are three stages involved in this—as there are in tackling of most problems—investigation (surveys), immediate action (control) and future planning (design). These are applicable not only to the overall problems of traffic policy but also to the individual traffic problems with which the majority of traffic engineers will be concerned.

Before we can start to alleviate a particular traffic problem—for example, a whole town, a busy road, or just one congested intersection—we need a detailed survey of the traffic, and we need to be able to interpret the information thus obtained. Traffic engineering employs certain survey techniques and statistical methods of interpretation for these purposes.

Having investigated a problem in detail, it is seldom possible to rebuild or re-align a road at short notice, and in fact this may not be necessary. But it is often essential to alleviate the problem by some immediate treatment, either by management techniques or by control of the traffic. As we have already remarked, a tin of white paint may be all that is needed, or an adjustment to a traffic signal setting. Some immediate solutions however are only valid in the short term—we must also look to the future and design road networks, intersections and new towns to cope with the foreseeable future traffic demand.

In the past, decisions on road alignments, road widths, intersection design, etc. were based on "rule-of-thumb" methods derived from long experience. In many cases the decision were correct—but not always. With the continuing growth of the traffic problem we cannot afford to design "off-the-cuff". It is for this reason that modern traffic engineering theories and techniques have been developed, based on scientific appraisal of proven experience, particularly in Britain and America.

It will be appreciated that the problems of traffic engineering are most pressing in urban areas. We shall concentrate on dealing with urban traffic problems and techniques, looking at rural problems on passing. And the traffic engineer should be checking to ensure that he is at least not damaging the environment, and is preferably improving it. Traffic engineering today, more than ever, is about improving the quality of life.

Summary

(1) Traffic engineering is concerned with the planning, functional design and operation of roads and their associated transport facilities, while maintaining or improving the environmental quality of life.

(2) Car ownership has grown rapidly in the past. It continues to grow—the car is a world wide status symbol.

(3) Sufficient roads cannot be provided to accommodate the demand.

(4) The best way the growth in car ownership can be accommodated is by combining better use of existing facilities, some new facilities, and some restraint in the use of the car.

(5) The way to devise such a solution is by a logical process of survey, immediate control and future design—of roads, intersections and car-parks.

Words and Expressions

derivation   n. 引出,衍生
amenity   n. (环境、气候的)舒服
alignment   n. 定线
traffic stream   车流,交通流
temporary   a. 暂时的,临时的
capital investment   投资
bypass   n. (为避免交通车辆拥挤而筑的)迂回的旁道
alleviate   v. 减轻,缓和
interpret   v. 解释,说明
rule-of-thumb   经验的
off-the-cuff   (俚)即兴的,非正式的
appraisal   n. 评价

Note

①We can think of these words as embracing traffic flow, … with which aspects traffic economics is closely interwoven; and throughout, the concern for environmental compatibility. 定语从句 with which aspects… interwoven 修饰 traffic flow; and throughout, … compatibility(自始至终要关心环境的适合问题)补充说明前面,起状语作用。

UNIT 21

Text A Bridges

When designing and constructing a long-span bridge the great weight of the structure, the dynamic effects of moving loads such as locomotives or motor vehicles, and the aerodynamic effects of wind pressure give rise to problems which call for the greatest knowledge and ingenuity in their solution[①]. Not the least important of the problems concerns the construction of the foundations, particularly when these have to be laid on the beds of rivers.

The construction of a long-span bridge is a great achievement, and the history of bridge building includes many human romantic and even tragic stories.

All bridges may be broadly classified under three headings: beam (including the truss and cantilever), arch, and suspension.

Beam (Simple and Continuous) Bridges

A simple single-span bridge, the cantilever may be of steel (probably a plate girder), reinforced concrete or prestressed concrete. In steel the maximum span for a simple beam bridge is usually about 100 ft (although bridges with longer spans have been built). When, however, the spans are large, a continuous girder is usually adopted. A plate-girder bridge in Germany has a central span of 354 ft and side spans of 295 ft.

Beam (Truss) Bridges

For spans between supporting piers above about 150 ft, the truss often provides the most economical bridge and the material is invariably steel. A bridge over the Ohio in Illinois, completed in 1917, has a simply supported span of 720 ft.

Beam (Cantilever) Bridges

The principle of the cantilever bridge, with and without a suspended span, is illustrated in Fig. 1 (a) and (b), although the latter is not common in solid beam or girder construction. The piers having been built, the bridge (anchored at A and B) is built out from each pier, and the middle portion of the bridge, called the suspended span, which is usually in one prefabricated unit, is then placed in position[②]. The bridge therefore consists of two anchored cantilevers supporting [in type (a)] a beam "suspended" from the ends of the cantilevers. The maximum bending moments and shear forces occur at C and D, and at these points the bridge is usually of greater depth.

When spans are large, thus requiring a great depth of bridge, cantilever bridges are usually constructed of steel trusses (trussed girders). It is possible in this way to have spans of up to about 1,800 ft between piers. Fig. 1 (c) is an example where the cantilevers meet without a middle suspended span. Although this bridge may look like an arch, it is in fact a double-cantilever truss or trussed beam. It may be noted that in cantilever bridges the greatest depth of truss occurs at the main piers because it is at these points that the greatest stresses occur.

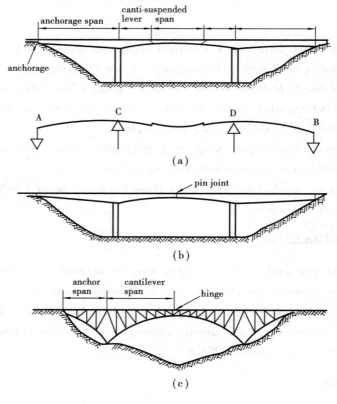

Fig. 1

Arch Bridges

In an arch bridge the arch is the main structural member and transmits the loads imposed on it to the abutments at the springing points. The part of the construction above the arch ring when the roadway or railway is at a higher level than the crown of the arch is called the spandrel[③].

Since steel and reinforced concrete are capable of taking tension, the arch rings can be very much thinner than in masonry construction. The braced spandrel bridge is usually constructed in steel, as is also the bridge where the roadway is supported by hangers from the structural arch.

Another type of arched bridge is the stiffened tied-arch, which is often called a bow-string girder. In an archery bow the string prevents the bow from flattening out. In a similar manner, the road-supporting horizontal girders are made strong enough to absorb the arch thrusts, and therefore the reactions on the piers and abutments are vertical.

Suspension Bridges

When spans are large, about 2,000 ft or more, suspension bridges are the most economical, but they can, of course, be used for smaller spans. Usually, there is a central span with two side spans and the cables passing over the top of the supporting piers are anchored in tunnel or by other means. Since the cables pull on each pier, the load on the pier is entirely or almost vertical. The roadway is suspended from the inclined cables by vertical hangers.

Words and Expressions

aerodynamic　　*a.* 空气动力学的
ingenuity　　*n.* 创造(独创)性,精巧
not the least　　很,非常
cantilever　　*n.* 悬臂,伸臂,悬臂梁
girder　　*n.* 大梁
plate girder　　板梁
central span　　中跨
side span　　边跨
pier　　*n.* 桥墩
truss　　*n.* 桁架
suspended span　　挂孔,吊梁
solid　　*a.* 实体的,固体的,坚实的;*n.* 实体,固体
anchor　　*v.* 固定
prefabricate　　*v.* 预制
bending moment　　弯矩
shear force　　剪力
depth of bridge　　桥梁建筑高度
trussed beam (girder)　　桁(架)梁
abutment　　*n.* 桥台,拱座
springing point　　开始点,起拱点

crown　*n.* 拱顶,路拱
spandrel　*n.* 拱肩,拱上空间
arch ring　拱圈
brace　*v.* 支撑,联结
braced arch　桁拱
hanger　*n.* 吊杆,吊架,钩子
tied-arch　系杆拱
bow-string girder　弓弦梁,弓弦式桁梁
archery bow　射箭用弓
flatten　*v.* 整平,压扁
thrust　*n.* 推力,侧向推力; *v.* 推,推入,冲入
suspension bridge　吊桥,悬索桥

Notes

①When designing and constructing along-span bridge the great weight of the structure, the dynamic effects of moving loads such as locomotives and motor vehicles, and the aerodynamic effects of wind pressure give rise to problems which call for the greatest knowledge and ingenuity in their solution.当设计和建造一座大跨径桥梁时,结构的巨大重量,机车和机动车辆等动荷载的动力作用以及风压的空气动力效应,常常引起一些需要具有广博的知识和创造力的人才能解决的问题。句中 which 引导限定性定语从句,说明主句中的宾语名词 problems,同时在从句中担任主语。

②The piers having been built, the bridge (anchored at A and B) is built out from each pier, and the middle portion of the bridge, called the suspended span, which is usually in one prefabricated unit, is then placed in position.桥墩建好后,桥(在 A 和 B 锚住)就由各桥墩修出,然后桥的中段安装就位;中段称挂孔,通常为单个预制件。

having been built 是独立分词结构,作状语。因其逻辑主语不是句子的主语 bridge,而是自己的主语 piers,故须在前加上 piers。独立分词结构的逻辑主语通常就是句子的主语,在前面不另加词。独立分词结构可以是分词或分词短语,位于句首或句末,用逗号分开。本句的独立分词结构中的分词是完成时态被动形式。

句中的 which 引导非限定性定语从句,说明 suspended span。非限定性定语从句前有逗号,这种从句与其所说明句的关系不很密切,只起附加说明作用,英译汉时,一般与主句分开。

③The part of the construction above the arch ring when the roadway or railway is at higher level than the crown of the arch is called the spandrel.当公路或铁路标高位于拱顶以上平面位置时,拱圈以上的结构部分称为拱上建筑。

Text B　The Future of Tall Building

Zoning effects on the density of tall building and solar design may raise ethical questions.

Energy limitations will continue to be a unique design challenge. A combined project of old and new buildings may bring back human scale to our cities. Owners and conceptual designers will be challenged in the 1980s to produce economically sound, people-oriented buildings.

In 1980 the Lever House, designed by Skidmore, Owings and Merrill(SOM) received the 25-year award from the American Institute of Architects "in recognition of architectural design of enduring significance". This award is given once a year for a building between 25 and 35 years old. Lewis Mumford described the Lever House as "the first office building in which modern materials, modern construction, modern functions have been combined with a modern plan". At the time, this daring concept could only be achieved by visionary men like Gordon Bunshaft, the designer, and Charles Luckman, the owner and then-president of Lever Brothers. The project also included a few "first": (1) it was the first sealed glass tower ever built; (2) it was the first office building designed by SOM; and (3) it was the first office building on Park Avenue to omit retail space on the first floor. Today, after hundreds of look-alikes and variations on the grid design, we have reached what may be the epitome of tall building design: the nondescript building. Except for a few recently completed buildings that seem to be people-oriented in their lower floors, most tall buildings seem to be a repetition of the dull, graph-paper-like monoliths in many of our cities. Can this be the end of the design-line for tall building? Probably not. There are definite signs that are most encouraging. Architects and owners have recently begun to discuss the design problem publicly. Perhaps we are at the threshold of a new era. The 1980s may bring forth some new visionaries like Bunshaft and Luckman. If so, what kinds of restrictions or challenges do they face?

Zoning

Indications are strong that cities may restrict the density of tall buildings, that is, reduce the number of tall buildings per square mile. In 1980 the term grid-lock was used for the first time publicly in New York City. It caused a terror-like sensation in the pit of one's stomach. The term refers to a situation in which traffic comes to a standstill for many city blocks in all directions. The jam-up may even reach to the tunnels and bridges. Strangely enough, such an event happened in New York in a year of fuel shortages and high gasoline prices. If we are to avoid similar occurrences, it is obvious that the density of people, places, and vehicles must be drastically reduced. Zoning may be the only long-term solution.

Solar zoning may become more and more popular as city residents are blocked from the sun by tall buildings. Regardless of how effectively a tall building is designed to conserve energy, it may at the same time deprive a resident or neighbor of solar access. In the 1980s the right to see the sun may become a most interesting ethical question that may revolutionize the architectural fabric of the city. Mixed-use zoning which became a financially viable alternative during the 1970s, may become commonplace during the 1980s, especially if it is combined with solar zoning to provide access to the sun for all occupants.

Renovation

Emery Roth and Sons designed the Palace Hotel in New York as an addition to a renovated historic Villard house on Madison Avenue. It is a striking example of what can be done with salvageable and beautifully detailed old buildings. Recycling both large and small buildings may become the way in which humanism and warmth will be returned to buildings during the 80's. If we must continue to design with glass and aluminum in stark grid patterns, for whatever reason, we may find that a combination of new and old will become the great humane design trend of the future.

Conceptual Design

It has been suggested in architectural magazines that the Bank of America office building in San Francisco is too large for the city's scale. It has also been suggested that the John Hancock Center in Boston is not only out of scale but also out of character with the city. Similar statements and opinions have been made about other significant tall buildings in cities throughout the world. These comments raise some basic questions about the design process and who really makes the design decisions on important structures—and about who will make these decisions in the 1980s.

Will the forthcoming visionaries-architects and owners-return to more humane designs?

Will the sociologist or psychologist play a more important role in the years ahead to help convince these visionaries that a new, radically different, human-scaled architecture is long overdue? If these are valid questions, could it be that our "best" architectural designers of the 60's and 70's will become the worst designers of the 80's and 90's? Or will they learn and respond to a valuable lesson they should have learned in their "History of Architecture" course in college that "architecture usually reflects the success or failure of a civilized society"? Only time will tell.

Words and Expressions

zone   n. 区域
ethical   a. 道德的, 伦理的
challenge   n. 挑战
award   n. 奖, 奖状, 奖品
visionary   a. 幻觉的, 梦幻的, 空想的, 不实际的
retail   n. & v. 零售, 零卖
grid   n. 柱网
epitome   n. 缩影
nondescript   a. (因无特征而)难以归类的, 难以形容的
monolith   n. (柱状或碑状的)独块巨石, 独石柱
threshold   n. 开端; 门槛
standstill   n. 停止, 停滞不前
jam-up   交通堵塞

conserve   v. 节约
deprive   v. 剥夺
viable   a. 可行的；能生存的
renovate   v. 革新，更新；修复，修理
commonplace   a. 平凡的；陈腐的
occupant   n. 居住者；占用者
salvageable   a. 可救者，可抢救的
humanism   n. 人情，人性
stark   a. 僵硬的，刻板的；轮廓明显的
forthcoming   a. 即将到来的，即将出现的
radically   a. 根本的，基本的
overdue   a. 期待已久的，早就成熟的

UNIT 22

Text A   Civil Engineering Contracts

A simple contract consists of an agreement entered into by two or more parties, whereby[①] one of the parties undertakes to do something in return for something to be undertaken by the other. A contract has been defined as an agreement which directly creates and contemplates an obligation. The word is derived from the Latin *contractum*, meaning drawn together.

We all enter into contracts almost every day for the supply of goods, transportation and similar services, and in all these instances we are quite willing to pay for the services we receive. Our needs in these cases are comparatively simple and we do not need to enter into lengthy or complicated negotiations and no written contract is normally executed. Nevertheless, each party to the contract has agreed to do something, and is liable for breach of contract if he fails to perform his part of the agreement.

In general, English law requires no special formalities in making contracts but, for various reasons, some contracts must be made in a particular form to be enforceable and, if they are not made in that special way, then they will be ineffective. Notable among these contracts are contracts for the sale and disposal of land, and "land", for this purpose, includes anything built on the land, as, for example, roads, bridges and other structures.

It is sufficient in order to create a legally binding contract, if the parties express their agreement and intention to enter into such a contract. If, however, there is no written agreement and a dispute arises in respect of the contract, then the Court that decides the dispute will need to ascertain the terms of the contract from the evidence given by the parties, before it can make a decision on the matters in dispute.

On the other hand if the contract terms are set out in writing in a document, which the parties subsequently sign, then both parties are bound by these terms even if they do not read them. Once a person has signed a document he is assumed to have read and approved its contents, and will not be able to argue that the document fails to set out correctly the obligations which he actually agreed to perform. Thus by setting down the terms of a contract in writing one secures the double advantage of affording evidence and avoiding disputes.

The law relating to contracts imposes on each party to a contract a legal obligation to perform or observe the terms of the contract, and gives to the other party the right to enforce the fulfillment of these terms or to claim "damages" in respect of the loss sustained in consequence of the breach of contract.

Most contracts entered into between civil engineering contractors and their employers are of the type known as "entire" contracts. These are contracts in which the agreement is for specific works to be undertaken by the contractor and no payment is due until the work is complete.

In an entire contract, where the employer agrees to pay a certain sum in return for civil engineering work, which is to be executed by the contractor, the contractor is not entitled to any payment if he abandons the work prior to completion, and will be liable in damages for breach of contract. Where the work is abandoned at the request of the employer, or results from circumstances that were clearly foreseen when the contract was entered into and provided for in its terms, then the contractor will be paid as much as he has earned.

It is, accordingly, in the employer's interest that all contracts for civil engineering work should be entire contracts to avoid the possibility of work being abandoned prior to completion. However, contractors are usually unwilling to enter into any contracts, other than the very smallest, unless provision is made for interim payments to them as the work proceeds. For this reason the standard form of civil engineering contract provides for the issue of interim certificates at various stages of the works.

It is customary for the contract further to provide that a prescribed proportion of the sum due to the contractor on the issue of a certificate shall be withheld. This sum is known as "retention money" and serves to insure the employer against any defects that may arise in the work. The contract does, however, remain an entire contract, and the contractor is not entitled to receive payment in full until the work is satisfactorily completed, the maintenance period expired and the maintenance certificate issued.

That works must be completed to the satisfaction of the employer, or his representative[2], does not give the employer the right to demand an unusually high standard of quality throughout the works, in the absence of a prior express agreement. Otherwise the employer might be able to postpone indefinitely his liability to pay for the works. The employer is normally only entitled to expect a standard of work that would be regarded as reasonable by competent persons with considerable experience in the class of work covered by the particular contract. The detailed requirements of the specification will have a considerable bearing on these matters.

The employer or promoter of civil engineering works normally determines the conditions of contract, which define the obligations and performances to which the contractor will be subject. He often selects the contractor for the project by some form of competitive tendering and any contractor who submits a successful tender and subsequently enters into a contract is deemed in law to have voluntarily accepted the conditions of contract adopted by the promoter.

The obligations that a contractor accepts when he submits a tender are determined by the form of the invitation to tender. In most cases the tender may be withdrawn at any time until it has been accepted and may, even then, be withdrawn if the acceptance is stated by the promoter to be "subject to formal contract" as is often the case.

The employer does not usually bind himself to accept the lowest or indeed any tender and this is often stated in the advertisement[③]. A tender is, however, normally required to be a definite offer and acceptance of it gives rise legally to a binding contract.

A variety of contractual arrangements are available and the engineer will often need to carefully select the form of contract which is best suited for the particular project. The employer is entitled to know the reasoning underlying the engineer's choice of contract.

Types of contract are virtually classified by their payment system:

(1) pricebased: lump sum and admeasurement (prices or rates are submitted by the contractor in his tender);

(2) cost-based: cost-reimbursable and target cost (the actual costs incurred by the contractor are reimbursed, together with a fee for overheads and profit).

Words and Expressions

contract $n.$ 合同,契约
inter into 参加,受……约束
in return for 作为……的报酬
contemplate $v.$ 期望
obligation $n.$ 义务,责任
execute $v.$ 实施,贯彻
breach $n.$ 违反,不履行
breach of contract 违约
formality $n.$ 拘泥形式;[复]正式手续
disposal $n.$ 处理;卖掉,让与
binding contract 有(法律)约束力的合同
dispute $n.$ & $v.$ 争执,争端
ascertain $v.$ 查清,确定
contract term 合同条款
damage $n.$ 损坏,毁坏;[复]赔偿金
sustain $v.$ 蒙受,遭受
contractor $n.$ 承包(商)人
employer $n.$ 雇主,使用者,业主
entitle $v.$ 给……权利(资格)
interim $n.$ & $a.$ 中间(的),临时的

interim certificate   中间验收,临时证书
withhold   v. 拒绝,抑制
retention   n. 保留,记忆力
retention money   保留金
expire   v. 期满,终止
liability   n. 义务,法律责任
promoter   n. 发包人,发起人
tender   v. & n. 招(股)标,标书
deem   v. 认为,相信
virtually   ad. 实质上,事实上
reimburse   v. 赔偿,偿还
lump sum contract   总价合同
admeasurement contract   计价合同
cost-reimbursable contract   成本补偿合同
target cost contract   目标成本合同
incur   v. 招致
overhead   n. 管理费

Notes

①whereby = by which,意思是"借此",引出一个非限定性定语从句。

②that works must be completed to the satisfaction of the employer, or his representative,... 由that 引导的主语从句。

③本句可译为:雇主一般会说明他自己不受合同报价最低的投标者或任何投标者的约束,这一点在投标者之中往往已经声明。

Text B Making a Contract

The business man concerned with contracts will normally learn about their contents, and the precise but intricate mode of expression that is common to them and to all legal documents, in his own language. Thus, when he comes to tackle a contract written in English, his main difficulties will be understanding the meaning of the language, or choosing the appropriate expressions when drawing up an English-language agreement himself.

This article sets out to give a selection of the vocabulary commonly found in contracts. It is largely drawn from the General Conditions concerning contracts prepared by the United Nations Economic Commission for Europe.

Formation of the Contract

The parties to the contract should state when the contract is considered to become binding. A clause should be included stating something like this: "The Contract shall be deemed to have been entered into when, upon receipt of an order, the Vendor has sent an acceptance in writing within the time-limit (if any) fixed by the Purchaser."

It may be necessary to add that where an expert or import licence or a foreign exchange control authorization is required for the performance of the contract, the party responsible for obtaining the licence or authorization should make sure to obtain it in good time.

Terms of Delivery

The delivery period should be stated. It will often run from the date of the formation of the contract, or the date of the receipt by the Vendor of such payment in advance of delivery as is stipulated. On expiry of the delivery period provided for in the contract, the Vendor may be entitled to an additional period of grace.

There will probably be a clause concerning delay in delivery, granting a reasonable extension of the delivery period. Should the Vendor fail to deliver the goods after the period or grace, the Purchaser is normally entitled to terminate the contract by notice in writing to the Vendor. The Purchaser will then be entitled to recover any payment he has made in respect of undelivered goods, and to reject goods delivered which are unusable.

Questions of Payment

The manner of payment and the time at which it should be made must naturally be agreed upon. The time may, for instance, be thirty days after notification from the Vendor that the goods have been placed at the Purchaser's disposal.

If there is delay in payment, the Vendor may postpone the fulfillment of his own obligations until payment is made, or recover interest on the sum due, 6% is a common rate in such cases.

Guarantee

The contract will generally contain a guarantee, according to which the Vendor may undertake to remedy any defect resulting from faulty design, materials or workmanship. Certain limitations to the Vendor's liability for defects will be stated.

If, when the goods are inspected, it is found that they do not conform with the contract, the Purchaser will be entitled to reject the goods.

Arbitration

Since it may sometimes be difficult for the parties to the contract to settle by agreement some dispute arising out of or in connection with the contract, it is necessary to state an arbitrator who can settle such disputes out of court. Unless otherwise agreed, contracts are commonly governed by the law of the Vendor's country.

Some General Points

Other matters to be considered when drawing up a contract are packing, inspection and tests. In the case of plant and heavy machinery, there will be drawing and descriptive documents, and such questions as working conditions and safety regulations while erecting the plant.

Words and Expressions

intricate   *a.* 复杂的,难懂的
clause   *n.* 条款,款项
order   *n.* 订货单
vendor   *n.* 卖主
purchaser   *n.* 买主
foreign exchange   外汇兑换
authorization   *n.* 认可,授权
stipulate   *v.* 规定,约定
expiry   *n.* 期满,终止
period of grace   宽限日期,优惠期
terminate   *v.* 终止,结束
recover   *v.* 弥补,挽回
liability   *n.* 责任,义务
conform   *v.* 符合
arbitration   *n.* 仲裁
commission   *n.* 代表最高权力的委员会
binding   *a.* 有约束力的,有束缚力的
erect   *v.* 安装,装配

UNIT 23

Text A Construction and Building Inspectors

About half of all inspectors worked for local governments, primarily municipal or county building departments.

Opportunities should be best for experienced construction supervisors and craftworkers who have some college education, engineering or architectural training, or certification as construction inspectors or plan examiners.

Nature of the Work

Construction and building inspectors examine the construction, alteration, or repair of buildings, highways and streets, sewer and water systems, dams, bridges, and other structures to ensure compliance with building codes and ordinances, zoning regulations, and contract specifications. Building codes and standards are the primary means by which building construction is regulated in the United States to assure the health and safety of the general public. Inspectors make an initial inspection during the first phase of construction, and follow-up inspections throughout the construction project to monitor compliance with regulations[①]. However, no inspection is ever exactly the same. In areas where certain types of severe weather or natural disasters are more common, inspectors monitor compliance with additional safety regulations designed to protect structures and occupants during these events.

Building Inspectors

Inspect the structural quality and general safety of buildings. Some specialize in such areas as structural steel or reinforced concrete structures. Before construction begins, plan examiners determine whether the plans for the building or other structure comply with building code regulations and if they are suited in the engineering and environmental demands of the building site. Inspectors visit the worksite before the foundation is poured to inspect the soil condition and positioning and depth of the footings. Later, they return to the site to inspect the foundation after it has been completed. The size and type of structure, as well as the rate of completion, determine the number of other site visits they must make. Upon completion of the project, they make a final comprehensive inspection.

In addition to structural characteristics, a primary concern of building inspectors is fire safety. They inspect structures' fire sprinklers, alarms, and smoke control systems, as well as fire exits. Inspectors assess the type of construction, building contents, adequacy of fire protection equipment, and risks posed by adjoining buildings.

Electrical Inspectors

There are many types of inspectors. Electrical inspectors examine the installation of electrical systems and equipment to ensure that they function properly and comply with electrical codes and standards. They visit worksites to inspect new and existing sound and security systems, wiring, lighting, motors, and generating equipment. They also inspect the installation of the electrical wiring for heating and air conditioning systems, appliances, and other components.

Elevator Inspectors

Elevator inspectors examine lifting and conveying devices such as elevators, escalators, moving sidewalks, lifts and hoists, inclined railways, ski lifts, and amusement rides.

Mechanical Inspectors

Mechanical inspectors inspect the installation of the mechanical components of commercial kitchen appliances, heating and air-conditioning equipment, gasoline and butane tanks, gas and oil piping, and gas-fired and oil-fired appliances. Some specialize in boilers or ventilating equipment as well.

Plumbing Inspectors

Plumbing inspectors examine plumbing systems, including private disposal systems, water supply and distribution systems, plumbing fixtures and traps, and drain, waste, and vent lines.

Public Works Inspectors

Public works inspectors ensure that Federal, State, and local government construction of water and sewer systems, highways, streets, bridges, and dams conforms to detailed contract specifications. They inspect excavation and fill operations, the placement of forms for concrete, concrete mixing and pouring, asphalt paving, and grading operations. They record the work and materials used so that contract payments can be calculated. Public works inspectors may specialize in highways, structural steel, reinforced concrete, or ditches. Others specialize in dredging operations required for bridges and dams or for harbors.

Home Inspectors

Home inspectors generally conduct inspections of newly built or previously owned homes. Increasingly, prospective home buyers hire home inspectors to inspect and report the condition of a home's systems, components, and structure. They typically are hired either immediately prior to a purchase offer on a home, or as a contingency to a sales contract. In addition to structural quality, home inspectors must be able to inspect all home systems and features, from plumbing, electrical, and heating or cooling systems to roofing.

The owner of a building or structure under construction employs specification inspectors to ensure that work is done according to design specifications②. They represent the owner's interests, not those of the general public. Insurance companies and financial institutions also may use specification inspectors.

Details concerning construction projects, building and occupancy permits, and other documentation generally are stored on computers so that they can easily he retrieved, kept accurate, and updated. For example, inspectors may use laptop computers to record their findings while inspecting a site. Most inspectors use computers to help them monitor the status of construction inspection activities and keep track of issued permits.

Although inspections are primarily visual, inspectors may use tape measures, survey instruments, metering devices, and test equipment such as concrete strength measurers. They keep a log of their work, take photographs, file reports, and, if necessary, act on their findings. For example, construction inspectors notify the construction contractor, superintendent, or supervisor when they discover a code or ordinance violation or something that does not comply with the contract specifications or approved plans. If the problem is not corrected within a reasonable or specified period, government inspectors have authority to issue a "stop-work" order.

Many inspectors also investigate construction or alterations being done without proper permits. Inspectors who are employees of municipalities enforce laws pertaining to the proper design, construction, and use of buildings③. They direct violators of permit laws to obtain permits and submit to inspection.

Words and Expressions

inspector   *n.* 检查员,巡视员
building inspector   房屋监工员,监理工程师
home inspector   住宅检测员
municipal   *a.* 市政的,市立的;地方性的,地方自治的
compliance with   依从,符合
supervisor   *n.* 监督人,管理人,检查员,督学,主管人
construction inspector   工程检查员,建筑师驻现场代表
construction supervisor   施工管理员

examiner　　*n.*　观测员,检查员
craftworker　　*n.*　工匠,技工
under construction　　在建筑(施工)中,正在施工
building permit　　建筑许可证
occupancy permit　　居住证,占用许可证;使用执照
certification　　*n.*　证明
follow-up　　继续的,作为重复的
sewer　　*n.*　排水,污水
code　　*n.*　代码,代号,密码,编码
building code　　建筑法规,建筑规范
ordinance　　*n.*　法令,训令,布告,条例
building ordinance　　建筑法令
zoning　　*n.*　分区制
zoning regulation　　分区条例(规章,法规)
contract specification　　合同说明,合同说明书,施工说明书
monitor　　*v.*　监视,检验
positioning　　*n.*　配置,布置
comprehensive　　*a.*　全面的,广泛的
adjoining　　*a.*　邻接的,隔壁的
generate equipment　　发电设备
appliance　　*n.*　装置,器械
gas-fired appliance　　煤气用具
oil-fired appliance　　燃油用具
sidewalk　　*n.*　人行道
hoist　　*n.*　提升间,升起
ski lift　　滑雪用提升机
butane　　*n.*　丁烷
ventilate　　*v.*　通风
fixture　　*n.*　固定器械
plumbing fixture　　卫生设备,卫生器具,市内管理工程固定装置
trap　　*n.*　存水弯
distribution system　　配水(电)系统
dredge　　*v.*　疏浚(河道、港湾等),挖掘(泥土等)
contingency　　*n.*　意外事故,偶发事件;可能性,偶然性;意外开支
retrieve　　*n.*　重新得到,找回
laptop computer　　*n.*　便携式电脑
keep track of　　明了
log　　*n.*　日志;*v.*　把……记入日志
notify　　*n.*　通报

superintendent　　*n.* 主管,负责人,指挥者,管理者
municipality　　*n.* 市政当局,自治市
enforce　　*v.* 强迫,执行,坚持,坚强
pertain to　　适合,相称

Notes

①Inspectors make an initial inspection during the first phase of construction, and follow-up inspections throughout the construction project to monitor compliance with regulations. 本句可译为:监理人员在施工第一阶段即开始初始监理,并在整个项目的建造过程中继续监理,以控制施工过程,使其符合规范要求。本句主语为:Inspectors,谓语为 make,宾语有两个:一是 an initial inspection,后跟时间状语 during the first phase of construction;一是 follow-up inspections,后跟时间状语 throughout the construction project。To monitor compliance with regulations 为目的状语。

②The owner of a building or structure under construction employs specification inspectors to ensure that work is done according to design specifications. 在建的建筑物或构筑物的业主,雇用能执行规范的监理人员,以保证工作是按设计说明书进行的。句中主语为 The owner,谓语为 employs,to ensure that 为目的状语。

③Inspectors who are employees of municipalities enforce laws pertaining to the proper design, construction, and use of buildings. 受雇于市政当局的监理人员需依照有关设计、建造、建筑物使用与要求严格执法。本句主语为 Inspectors,谓语为 enforce laws pertaining to ...。who are employees of municipalities 为修饰 Inspectors 的定语从句。

Text B　　Construction Cost Estimation

Cost estimating is one of the most important steps in project management. A cost estimate establishes the base line of the project cost at different stages of development of the project. A cost estimate at a given stage of project development represents a prediction provided by the cost engineer or estimator on the basis of available data. According to the American Association of Cost Engineers, cost engineering is defined as that area of engineering practice where engineering judgment and experience are utilized in the application of scientific principles and techniques to the problem of cost estimation, cost control and profitability①.

Virtually all cost estimation is performed according to one or some combination of the following basic approaches.

Production Function

In microeconomics, the relationship between the output of a process and the necessary resources is referred to as the production function. In construction, the production function may be expressed by the relationship between the volume of construction and a factor of production such as labor or capital. A production function relates the amount or volume of output to the various inputs of labor, material and equipment. For example, the amount of output Q may be derived as a function of various input factors $x_1, x_2, \ldots, x_n$ by means of mathematical and/or statistical methods. Thus, for a specified level of output, we may attempt to find a set of values for the input factors so as to minimize the production cost. The relationship between the size of a building project (expressed in square feet) to the input labor (expressed in labor hours per square foot) is an example of a production function for construction.

Empirical Cost Inference

Empirical estimation of cost functions requires statistical techniques which relate the cost of constructing or operating a facility to a few important characteristics or attributes of the system[2]. The role of statistical inference is to estimate the best parameter values or constants in an assumed cost function. Usually, this is accomplished by means of regression analysis techniques.

Unit Costs for Bill of Quantifies

A unit cost is assigned to each of the facility components or tasks as represented by the bill of quantities. The total cost is the summation of the products of the quantities multiplied by the corresponding unit costs. The unit cost method is straightforward in principle but quite laborious in application. The initial step is to break down or disaggregate a process into a number of tasks. Collectively, these tasks must be completed for the construction of a facility. Once these tasks are defined and quantities representing these tasks are assessed, a unit cost is assigned to each and then the total cost is determined by summing the costs incurred in each task[3]. The level of detail in decomposing into tasks will vary considerably from one estimate to another.

Allocation of Joint Costs

Allocations of cost from existing accounts may be used to develop a cost function of an operation. The basic idea in this method is that each expenditure item can be assigned to particular characteristics of the operation. Ideally, the allocation of joint costs should be causally related to the category of basic costs in an allocation process. In many instances, however, a causal relationship between the allocation factor and the cost item cannot be identified or may not exist. For example, in construction projects, the accounts for basic costs may be classified according to (1) labor, (2) material, (3) construction equipment, (4) construction supervision, and (5) general office overhead. These basic costs may then be allocated proportionally to various tasks which are subdivisions of a project.

Construction cost constitutes only a fraction, though a substantial fraction, of the total project cost. However, it is the part of the cost under the control of the construction project manager. The required levels of accuracy of construction cost estimates vary at different stages of project development, ranging from ball park figures in the early stage to fairly reliable figures for budget control prior to construction[4]. Since design decisions made at the beginning stage of a project life cycle are more tentative than those made at a later stage, the cost estimates made at the earlier stage are expected to be less accurate. Generally, the accuracy of a cost estimate will reflect the information available at the time of estimation.

Construction cost estimates may be viewed from different perspectives because of different institutional requirements. In spite of the many types of cost estimates used at different stages of a project, cost estimates can best be classified into three major categories according to their functions. A construction cost estimate serves one of the three basic functions: design, bid and control. For establishing the financing of a project, either a design estimate or a bid estimate is used.

1. **Design estimates** For the owner or its designated design professionals, cost estimates encountered run parallel with the planning and design as follows:

- Screening estimates (or order of magnitude estimates).
- Preliminary estimates (or conceptual estimates).
- Detailed estimates (or definitive estimates).
- Engineer's estimates based on plans and specifications.

For each of these different estimates, the amount of design typically increases.

2. **Bid estimates** For the contractor, a bid estimate submitted to the owner either for competitive bidding or negotiation consists of direct construction cost including field supervision, plus a markup to cover general overhead and profits[5]. The direct cost of construction for bid estimates is usually derived from a combination of the following approaches.

- Subcontractor quotations.
- Quantity takeoffs.
- Construction procedures.

3. **Control estimates** For monitoring the project estimate is derived from available information to establish:

- Budget estimate for financing.
- Budgeted cost after contracting but prior to construction.
- Estimated cost to completion during the progress of construction.

Design Estimates

In the planning and design stages of a project, various design estimates reflect the progress of the design. At the very early stage, the screening estimate or order of magnitude estimate is usually made before the facility is designed, and must therefore rely on the cost data of similar facilities built in the past. A preliminary estimate or conceptual estimate is based on the conceptual design of the facility at the state when the basic technologies for the design are known. The detailed estimate or definitive estimate is made when the scope of work is clearly defined and the detailed design is in progress, so that the essential features of the facility are identifiable. The engineer's estimate is based on the completed plans and specifications when they are ready for the owner to solicit bids from construction contractors. In preparing these estimates, the design professional will include expected amounts for contractors' overhead and profits.

The costs associated with a facility may be decomposed into a hierarchy of levels that are appropriate for the purpose of cost estimation. The level of detail in decomposing the facility into tasks depends on the type of cost estimate to be prepared. For conceptual estimates, for example, the level of detail in defining tasks is quite coarse; for detailed estimates, the level of detail can be quite fine.

As an example, consider the cost estimates for a proposed bridge across a river. A screening estimate is made for each of the potential alternatives, such as a tied arch bridge or a cantilever truss bridge. As the bridge type is selected, e.g. the technology is chosen to be a tied arch bridge instead of some now bridge form, a preliminary estimate is made on the basis of the layout of the selected bridge form on the basis of the preliminary or conceptual design[6]. When the detailed design has progressed to a point when the essential details are known, a detailed estimate is made on the basis of the well defined scope of the project. When the detailed plans and specifications are completed, an engineer's estimate can be made on the basis of items and quantities of work.

Bid Estimates

The contractor's bid estimates often reflect the desire of the contractor to secure the job as well as the estimating tools at its disposal. Some contractors have well established cost estimating procedures while others do not. Since only the lowest bidder will be the winner of the contract in most bidding contests, any effort devoted to cost estimating is a loss to the contractor who is not a successful bidder. Consequently, the contractor may put in the least amount of possible effort for making a cost estimate if it believes that its chance of success is not high.

If a general contractor intends to use subcontractors in the construction of a facility, it may solicit price quotations for various tasks to be subcontracted to specialty subcontractors. Thus, the general subcontractor will shift the burden of cost estimating to subcontractors. If all or part of the construction is to be undertaken by the general contractor, a bid estimate may be prepared on the basis of the quantity takeoffs from the plans provided by the owner or on the basis of the construction procedures devised by the contractor for implementing the project[7]. For example, the cost of a footing of a certain type and size may be found in commercial publications on cost data which can be used to facilitate cost estimates from quantity takeoffs. However, the contractor may want to assess the actual coat of construction by considering the actual construction procedures to be used and the associated costs of the project is deemed to be different from typical designs[8]. Hence, items such as labor, material and equipment needed to perform various tasks may be used as parameters for the cost estimates.

Control Estimates

Both the owner and the contractor must adopt some base line for cost control during the construction. For the owner, a budget estimate must be adopted early enough for planning long term financing of the facility. Consequently, the detailed estimate is often used as the budget estimate since it is sufficient definitive to reflect the project scope and is available long before the engineer's estimate[9]. As the work progresses, the budgeted cost must be revised periodically to reflect the estimated cost to completion. A revised estimated cost is necessary either because of change orders initiated by the owner or due to unexpected cost overruns or savings.

For the contractor, the bid estimate is usually regarded as the budget estimate, which will be used for control purposes as well as for planning construction financing. The budgeted cost should also be updated periodically to reflect the estimated cost to completion as well as to insure adequate cash flows for the completion of the project.

Words and Expressions

estimate　　v. 估算,估计,估价
profitability　　n. 利益
inference　　n. 推理,推论,推断
parameter　　n. 参数
regression　　n. 回归,复归
regression analysis　　回归分析法
straightforward　　a. 坦率的,简单的,易懂的,直截了当的
disaggregate　　v. 使崩溃,分解,聚集
collectively　　ad. 全体地,共同地
incur　　v. 招致
category　　n. 类别
decompose　　v. & n. 分解,化解
joint cost　　联合费用
expenditure　　n. 消费,花费,使用
overhead　　n. 企业一般管理费,杂费,杂项开支,总开销
substantial　　a. 坚固的,实质的,真实的,充实的
ball park　　(数量、程度或质量)相近,大约
tentative　　a. 尝试的,暂时的
perspective　　n. 透视图,远景,前途,观点,看法,观察
bid　　v. 出价,投标,祝愿,命令,支付；　n. 出价,投标
designate　　n. 指明,指出,任命,指派；　v. 指定,指派
parallel　　a. 平行的,类似的；　n. 平行线(面),类似,相似物
run parallel with　　与……平行
markup　　n. 涨价,涨价幅度
competitive bidding　　竞标
quotation　　n. 报价单,估价单
takeoff　　n. 估计量
budget estimate　　概算
budgeted cost　　预算成本

conceptual　*a.* 概念的
identifiable　*a.* 可以确定的
solicit　*v.* 恳求
hierarchy　*n.* 层次,层级
tied arch bridge　系杆拱桥
cantilever truss bridge　悬臂桁架桥
subcontractor　*n.* 转包商,次承包商
base line　底线
budget control　预算控制,预算管理
revise　*v.* 修订,校订,修正,修改
initiate　*v.* 开始,发动,创始
overrun　*n.* 超支
building project　建筑工程
bill of quantifies　数量明细表,工程量清单
allocations of cost　工程费用分摊
estimated cost　预算价值(费用),估算成本
cash flow　(公司,政府等的)现金流转
break down　毁掉,制服,压倒,停顿,倒塌,中止,分解
be appropriate for　对……适合
at one's disposal　随某人自由处理,由某人随意支配

Notes

①According to the American Association of Cost Engineers, cost engineering is defined as that area of engineering practice where engineering judgment and experience are utilized in the application of scientific principles and techniques to the problem of cost estimation, cost control and profitability. 依照美国成本工程师协会的规定,造价工程即在运用科学原理和技术解决成本估算、成本控制和收益方面的问题时,需要运用工程判断和经验的工程实践领域。本句主语为 cost engineering, 谓语为 is defined; as that area of engineering practice 为主语补足语。where engineering profitability 为定语从句,修饰 that area。

②Empirical estimation of cost functions requires statistical techniques which relate the cost of constructing or operating a facility to a few important characteristics or attributes of the system. 成本函数的经验估算需要统计方法,将建造成本或开动设备的成本与系统的某些重要特征或属性联系起来。本句主语为 Empirical estimation of cost functions, 谓语为 requires, 宾语为 statistical techniques, which relate... to a few important characteristics or attributes of the system 为定语从句。

③Once these tasks are defined and quantities representing these tasks are assessed, a unit cost is assigned to each and then the total cost is determined by summing the costs incurred in each task. 一旦这些作业及其工程量确定下来,每个作业的单价便确定下来,然后将每个作业发生的费用加起来,就可确定总费用。本句中 Once these tasks are defined and quantities representing these tasks are assessed 为时间状语从句。

④The required levels of accuracy of construction cost estimates vary at different stages of project development, ranging from ball park figures in the early stage to fairly reliable figures for budget control prior to construction. 本句主语为 The required levels of accuracy of construction cost estimates,谓语为 vary,at different stages of project development 为时间状语,现在分词 ranging 的主语与句子的主语一致。本句可译为:在项目发展的不同阶段,建筑成本估算的精确程度各不相同,早期数字较为近似,而施工前预算控制时数字已相当准确。

⑤For the contractor, a bid estimate submitted to the owner either for competitive bidding or negotiation consists of direct construction cost including field supervision, plus a markup to cover general overhead and profits. 本句主语为 a bid estimate,谓语为 consists of,宾语为 direct construction cost,过去分词 submitted to the owner 为主语的定语,现在分词 including field supervision 作定语修饰 direct construction cost。本句可译为:对于承包人,为竞标或议标而递交给业主的投标报价由以下两部分组成:包括现场管理费用在内的直接施工成本及包含常规企业管理费用和利润在内的上涨幅度。

⑥As the bridge type is selected, e.g. the technology is chosen to be a tied arch bridge instead of some new bridge form, a preliminary estimate is made on the basis of the layout of the selected bridge form on the basis of the preliminary or conceptual design. 本句主句为被动语态,主语为 a preliminary estimate,谓语为 is made。As the bridge type is selected, e.g. the technology is chosen to be a tied arch bridge instead of some new bridge form 为条件状语从句。本句可译为:一旦桥梁类型确定后,例如,选定了系杆拱桥而不是其他新型桥梁,应以所选定桥型的设计图纸的初步设计或概念设计来进行设计概算。

⑦If all or part of the construction is to be under taken by the general contractor, a bid estimate may be prepared on the basis of the quantity takeoffs from the plans provided by the owner or on the basis of the construction procedures devised by the contractor for implementing the project. 本句主语为 a bid estimate,谓语为 maybe prepared。If all or part of the construction is to be undertaken by the general contractor. 为条件状语从句。本句可译为:如果总承包商承担全部或部分建造工作,投标报价将根据业主所提供的图纸或根据执行合同的承包商设计的施工程序来确定。

⑧However, the contractor may want to assess the actual cost of construction by considering the actual construction procedures to be used and the associated costs if the project is deemed to be different from typical designs. 本句主语为 the contractor,谓语为 may want,by considering the actual construction procedures to be used and the associated costs 为方式状语。If the project is deemed to be different from typical designs 为条件状语从句。本句可译为:但是,如果认为项目不同于典型的设计,承包商的实际建筑成本应结合实际的施工程序及其相关的费用来评定。

⑨Consequently, the detailed estimate is often used at the budget estimate since it is sufficient definitive to reflect the project scope and is available long before the engineer's estimate. 本句主语为 the detailed estimate,谓语为 is often used。since,…long before the engineer's estimate 为原因状语从句。本句可译为:由于施工图预算足以明确反映施工范围,并且早在施工预算之前就可以采用,因而,施工图预算经常用作概算。

UNIT 24

Text A Scientific Paper

The goal of scientific research is publications. Scientists, starting as graduate students, are measured primarily not by their dexterity in laboratory manipulations, not by their innate knowledge of either broad or narrow scientific subjects, and certainly not by their wit or charm; they are measured, and become known (or remain unknown), by their publications.

A scientific experiment, no matter how spectacular the results, is not completed until the results are published. In fact, the cornerstone of the philosophy of science is based on the fundamental assumption that original research must be published; only thus can new scientific knowledge be authenticated and then added to the existing data base that we call scientific knowledge.

It is not necessary for the plumber to write about pipes, nor is it necessary for the lawyer to write about cases (except brief writing), but the research scientist, perhaps uniquely among the trades and professions, must provide a written document showing what he or she has done, why it has been done, how it has been done, and what has been learned from it. The key word is reproducibility. That is what makes scientific writing unique.

Thus the scientist must not only "do" science but "write" science[①]. Bad writing can and often does prevent or delay the publication of good science. Unfortunately, the education of scientists is often so overwhelmingly committed to the technical aspects of science that the communication arts are neglected or ignored.

A scientific paper is a written and published report describing original research results. Therefore, a qualified short definition must be, given to the scientific paper. As defined by the tradition, editorial practices, scientific ethics, printing and publishing procedures that have developed over the past three centuries, a scientific paper must be written in a certain way and published in a certain way.

To properly define "scientific paper", we must define the mechanism that creates a scientific paper, namely, valid publication. Abstracts, theses, and conference literature, as well as institutional bulletins and other ephemeral publications, do not qualify as primary literature.

Many people have struggled with the definition of valid publication, from which is derived the definition of scientific paper.

An acceptable primary scientific publication must be the first disclosure containing sufficient information to enable peers (1) to assess observations, (2) to repeat experiments, and (3) to evaluate intellectual processes; moreover, it must be susceptible to sensory perception, essentially permanent, available to the scientific community without restriction, and available for regular screening by one or more of the major recognized secondary services (e. g., currently, The Engineering Index, Chemical Abstracts, Science Citation Index, etc., in the United States and similar services in other countries).

At first reading, this definition may seem excessively complex, or at least verbose. But those of us who had a hand in drafting it weighed each word carefully, and we doubt that an acceptable definition could be provided in appreciably fewer words. Because it is important that students, authors, editors, and all others concerned understand what a scientific paper is and what it is not, it may be helpful to work through this definition to see what it really means.

"An acceptable primary scientific publication" must be "the first disclosure". Certainly, first disclosure of new research data often takes place via oral presentation at a scientific meeting. But the thrust of the statement is that disclosure is more than disgorgement by the author; effective first disclosure is accomplished only when the disclosure takes a form that allows the peers of the author (either now or in the future) to fully comprehend and use that which is disclosed.

Thus, sufficient information must be presented so that potential users of the data can (1) assess observations, (2) repeat experiments, and (3) evaluate derivation processes (are the author's conclusions justified by the data?). Then, the disclosure must be "susceptible to sensory perception". This may seem an awkward phrase, because in normal practice it simply means publication; however, this definition provides for disclosure not just in terms of visual materials (printed journals, microfilm, microfiche) but also perhaps in nonprint, nonvisual forms. For example, "publication" in the form of audio cassettes, if that publication met the other tests provided in the definition, would constitute effective publication. In the future, it is quite possible that first disclosure will be entry into a computer data base.

Words and Expressions

  dexterity *n.* 灵巧,熟练,技巧,巧妙
  innate *a.* 先天的,固有的,内在的
  spectacular *a.* 壮观的,惊人的,引人注目的
  cornerstone *n.* 基石,柱石,基础
  authenticate *v.* 证明,证实,鉴定,辩证

plumber　　n. 管子工
reproducibility　　n. 可再现性,复验性,再生性,复现性
commit　　v. 把……交给
ethics　　n. 伦理观,道德标准,(某种职业的)规矩
interplay　　n. & v. 相互影响,相互作用
mechanism　　n. 机械,机理,历程,进程,技巧
thesis　　n. [复]theses 论文,论题
literature　　n. 文献
institutional　　a. 惯例的
ephemeral　　a. 生命短促的,瞬息的,短暂的
disclosure　　n. 揭发,泄露;公布,被揭发(公布)出来的事物
peer　　n. 同等,匹敌者,同行; v. 比得上
intellectual　　a. 智力的; n. 知识分子
susceptible　　a. 灵敏的,敏感的,容许的,被……的
sensory　　a. 感觉的,知觉的
perception　　n. 理解,感受,体会,观念,理解力
screen　　n. 屏幕,刷子; v. 屏幕,筛分,审查
verbose　　a. 啰唆的
appreciably　　ad. 相当地,可观地,显著地
work through　　看一遍,逐渐进行
disgorgement　　n. 吐出,流出,喷出,交出,讲出
microfilm　　n. 显微胶片,微缩胶卷
microfiche　　n. 缩微胶片,平片,显微照相卡片(4英寸×6英寸的卡片上有6帧×12帧面显微照相)

Note

①not only... but: 意思是:不仅做科学研究,而且还必须把研究成果写出来。

Text B　　How to Write a Scientific Paper

Title

　　In preparing a title for a paper, the author would do well to remember one salient fact: that title will be read by thousands of people. Perhaps few people, if any, will read the entire paper, but many people will read the title, either in the original journal or in one of the secondary (abstracting and indexing) services. Therefore, all words in the title should be chosen with great care, and their association with one another must be carefully managed.

The title of a paper is a label. It is not a sentence. Because it is not a sentence, with the usual subject, verb, object arrangement, it is really simpler than a sentence (or, at least, usually shorter), so the order of the words becomes even more important.

The meaning and order of the words in the title are of importance to the potential reader who sees the title in the journal table of contents. But these considerations are equally important to all potential users of the literature, including those (probably a majority) who become aware of the paper via secondary sources. Thus, the title should be useful as a label accompanying the paper itself, and it also should be in a form suitable for the machine-indexing systems used by Chemical Abstracts, The Engineering Index, Science Citation Index[1], and so on. Most of the indexing and abstracting services are geared to "key word" systems. Therefore, it is fundamentally important that the author provide the right "keys" to the paper when labeling it. That is to say, the terms in the title should be limited to those words that highlight the significant content of the paper in terms that are both understandable and retrievable.

Abstract

An Abstract should be viewed as a miniversion of the paper. The Abstract should provide a brief summary of each of the main sections of the paper. A well-prepared Abstract enables readers to identify the basic content of a document quickly and accurately, to determine its relevance to their interests, and thus to decide whether they need to read the document in its entirety. The Abstract should not exceed 250 words and should be designed to define clearly what is dealt with in the paper. Many people will read the Abstract, either in the original journal or in The Engineering Index, Science Citation Index, or one of the other secondary publications.

The Abstract should (1) state the principal objectives and scope of the investigation, (2) describe the methodology employed, (3) summarize the results, and (4) state the principal conclusions. The importance of the conclusions is indicated by the fact that they are often given three times: once in the Abstract, again in the Introduction, and again (in more detail probably) in the Discussion.

The Abstract should never give any information or conclusion that is not stated in the paper. References to the literature must not be cited in the Abstract (except in rare instances, such as modification of a previously published method).

Introduction

Now that we have finished the preliminaries, we come to the paper itself. I should mention that some experienced writers prepare their title and abstract after the paper is written, even though by placement these elements come first. You should, however, have in mind (if not on paper) a provisional title and an outline of the paper that you propose to write. You should also consider the level of the audience you are writing for, so that you will have a basis for determining which terms and procedures need definition or description and which do not.

The first section of the text proper[2] should, of course, be the Introduction. The purpose of the

Introduction should be to supply sufficient background information to allow the reader to understand and evaluate the results of the present study without needing to refer to previous publications on the topic. The Introduction should also provide the rationale for the present study. Above all, you should state briefly and clearly your purpose in writing the paper. Choose references carefully to provide the most important background information.

Suggested rules for a good Introduction are as follows: (1) It should present first, with all possible clarity, the nature and scope of the problem investigated. (2) It should review the pertinent literature to orient the reader. (3) It should state the method of the investigation. If deemed necessary, the reasons for the choice of a particular method should be stated. (4) It should state the principal results of the investigation. (5) It should state the principal conclusions suggested by the results. Do not keep the reader in suspense; let the reader follow the development of the evidence.

Materials and Methods

In the first section of the paper, the Introduction, you stated the methodology employed in the study. If necessary, you also defended the reasons for your choice of a particular method over competing methods.

Now, in the Materials and Methods, you must give the full details. The main purpose of the Materials and Methods section is to describe the experimental design and then provide enough detail that a competent peer can repeat the experiments. Many (probably most) readers of your paper will skip this section, because they already know (from the Introduction) the general methods you used and they probably have no interest in the experimental detail. However, careful writing of this section is critically important because the cornerstone of the scientific method requires that your results, to be of scientific merit, must be reproducible; and, for the results to be adjudged reproducible, you must provide the basis for repetition of the experiments for others. That experiments are unlikely to be reproduced is beside the point; the potential for producing the same or similar results must exist, or your paper does not represent good science.

When your paper is subjected to peer review, a good reviewer will read the Materials and Methods carefully. If there is a serious doubt that your experiments could be repeated, the reviewer will recommend rejection of your manuscript no matter how awe-inspiring your results are.

In describing the methods of the investigations, you should give sufficient details so that a competent peer could repeat the experiments, as stated above. If your method is new (unpublished) you must provide all of the needed details. However, if a method has been previously published in a standard journal, only the literature reference should be given.

Results

So now we come to the core of the paper, the data. This part of the paper is called the Results.

There are usually two components of the Results section. First, you should give an overall description of the experiments, providing the "big picture" without, however, repeating the experimental details previously provided in the Materials and Methods. Second, you should present the data.

Of course, it isn't quite that easy[③]. How do you present the data? A simple transfer of data from laboratory notebook to manuscript will hardly do. Most importantly, in the manuscript you should present representative data rather than endless repetitive data.

The Results needs to be clearly and simply stated, because it is the Results that comprise the new knowledge that you are contributing to the world. The earlier parts of the paper (Introduction, Materials and Methods) are designed to tell what they mean. Obviously, therefore, the whole paper must stand or fall on the basis of the Results. Thus, the Results must be presented with crystal clarity.

Discussion

The Discussion is harder to define than other sections. Thus, it is usually the hardest section to write. And, whether you know it or not, many papers are rejected by journal editors because of a faulty interesting. Even more likely, the true meaning of the data may be completely obscured by the interpretation presented in the Discussion, again resulting in rejection.

What are the essential features of a good Discussion? I believe the main components of the Discussion will be provided if the following suggestions are heeded:

(1) Try to present the principles, relationships, and generalizations shown by the Results. And bear in mind, in a good Discussion, you discuss—you do not recapitulate the Results.

(2) Point out any exception or any lacking of correlation and define unsettled points. Never take the high-risk alternative of trying to cover up or fudge data that do not quite fit.

(3) Show how your results and interpretations agree (or contrast) with previously published work.

(4) Don't be shy; discuss the theoretical implications of your work, as well as any possible practical applications.

(5) State your conclusions as clearly as possible.

(6) Summarize your evidence for each conclusion.

In showing the relationships among observed facts, you do not need to reach cosmic conclusions. Seldom will you be able to illuminate the whole truth; more often, the best you can do is to shine a spotlight on one area of the truth. Your one area of truth can be buttressed by your data; if you extrapolate to a bigger picture than that shown by your data, you may appear foolish to the point that even your data-supported conclusions are cast into doubt.

When you describe the meaning of your little bit of truth, do it simply. The simplest statements evoke the most wisdom; verbose language and fancy technical words are used to convey shallow thought.

Words and Expressions

salient   *a.* 突出的,显著的,优质的,明显的
index   [复]indexes 或 indices   *n.* 索引
table of contents   目录
gear   *v.* 使适合
highlight   *v.* 着重,使显著,使突出
retrievable   *a.* 可重新得到的,可恢复的
miniversion   *n.* 微缩版本
summary   *n.* 摘要,概要
identify   *v.* 识别,认出
methodology   *n.* 方法
preliminary   *a.* 初步的,序言的;*n.* [复]正文前面的内容,准备工作
placement   *n.* 放置,布置
provisional   *a.* 暂定的,假定的,暂时的,临时的
rationale   *n.* 基本原理,理论基础,理论的阐述
clarity   *n.* 清澈,明晰
pertinent   *a.* 相关的
orient   *v.* 定向,取向,正确地判断,(使)适应
deem   *v.* 认为,相信
suspense   *n.* 挂虑,不安,担心
merit   *n.* 价值
reproducible   *a.* 再现,复制
recommend   *v.* 建议
manuscript   *n.* 手稿
awe-inspiring   *a.* 使人畏惧的,使人敬畏的
core   *n.* 核心
comprise   *v.* 包含,构成
crystal   *a.* 水晶的,像水晶一样透明的
obscure   *v.* 搞混,使难理解;*a.* 模糊的,不清楚的,难解的
heed   *v.* 注意,留意
recapitulate   *v.* 概况,重现,再演
correlation   *n.* 关联,相关性,相互关系
unsettled   *a.* 不稳定的,不安定的,未解决的,混乱的
cover up   包裹,隐藏,掩盖
fudge   *n.* & *v.* 空话,捏造;粗制滥造,捏造
implication   *n.* 隐含,意义,本质
cosmic   *a.* 宇宙的,全世界的,广大无边的

illuminate   *v.* 照亮,阐明,使明白,使显扬,使光辉灿烂
buttress   *v.* & *n.* 支持,加强；支持物,支柱
extrapolate   *v.* 推断,外推,外插
evoke   *v.* 唤起,引起,博得
fancy   *n.* & *a.* 想象,美妙的,漂亮的

Notes

①Chemical Abstract,The Engineering Index,Science Citation Index:简写为 CA(化学摘录)、EI(工程索引)、SCI(科学检索)。

②proper:用在名词后面,意思是"本身的"。例如:the dictionary proper,表示"词典正文"。所以,本文中 the text proper,表示正文内容。

③that 是副词,修饰形容词 easy。

References

[1] 徐占发.建筑专业英语[M].北京:中国建材工业出版社,2003.
[2] 董亚明,佟方,庄思永.理工科专业英语[M].上海:华东理工大学出版社,2002.
[3] US Department of Transportation, Federal Highway Administration. Portland Cement Concrete Materials Manual [Z]. USA:Association Administrator for Engineering and Program Development, Office of Highway Operations, Construction and Maintenance Division, Materials Branch, 1990.
[4] 贾艳敏,施平.土木工程专业英语[M].2版.哈尔滨:哈尔滨工业大学出版社,2002.
[5] 周开鑫.土木类工程英语教程[M].北京:人民交通出版社,2002.
[6] 李嘉.专业英语(公路、桥梁工程专业用)[M].北京:人民交通出版社,1997.
[7] 周远棣.专业英语(公路与城市道路、桥梁工程专业用)[M].北京:人民交通出版社,1993.